Blockchain Technology for Business Processes

Katarina Adam

Blockchain Technology for Business Processes

Meaningful Use of the New Technology in Businesses

 Springer

Katarina Adam
Hochschule für Technik und Wirtschaft
Berlin, Germany

ISBN 978-3-662-65817-8   ISBN 978-3-662-65818-5   (eBook)
https://doi.org/10.1007/978-3-662-65818-5

Responsible Editor: Christine Sheppard
This Springer imprint is published by the registered company Springer-Verlag GmbH, DE, part of Springer Nature.
The registered company address is: Heidelberger Platz 3, 14197 Berlin, Germany

For my daugther Marleen

Contents

List of Abbreviations

ABS	Asset-Backed Securities
ARPA	Advanced Research Projects Agency
BaFin	Federal Financial Supervisory Authority
B2B	Business to Business
B2C	Business to Customer
BSI	Federal Office for Security and Information Technology
CBDC	Central Bank Digital Currency
DDoS	Distributed Denial of Service
DeFi	Decentralized Finance
DEX	Decentralized Exchange
DLT	Distributed Ledger Technology (is often used as an equivalent to the term blockchain.)
DNS	Domain Name System
ERC	Ethereum Request for Comments
EVM	Ethereum Virtual Machine
EZB	European Central Bank
GAN	Generative Adversarial Networks
GUI	Graphical User Interface
et seq.	Continued
ICO	Initial Coin Offering
IEO	Initial Exchange Offering
IFO	Initial Futures Offering
IPO	Initial Public Offering
IoT	Internet of Things
KAGB	German Investment Act (Kapitalanlagegesetzbuch)
KMU	small and medium-sized enterprises
KWG	German Banking Act
MAR	German Market Abuse Regulation
MIFID II	Market in Financial Instruments Directive II

NFT	Non Fungible Token
PoA	Proof of Authority
PoB	Proof of Burn
PoET	Proof of Elapsed Time
PoI	Proof of Importance
PoR	Proof of Reputation
POS	Proof of Stake
PoW	Proof of Work
P2P	Peer-to-Peer
SEC	Security and Exchange Commission
StGB	Strafgesetzbuch (Criminal Code)
VAG	Insurance Supervision Act
VDI	Association of German Engineers
VermAnlG	Investment Act
WpHG	German Securities Trading Act
WpPG	German Securities Prospectus Act
ZAG	German Payment Services Supervision Act
ZKP	Zero Knowledge Proof

Introduction

1

Abstract

The purpose of this book is to clarify misunderstandings about this technology, explain existing approaches, and empower the reader to make their own decisions about whether implementing a blockchain-based solution is worthwhile. Terms will be explained and placed in the correct context.

The target audience for this book are small and medium-sized enterprises (SMEs) that operate in value chains, for example as suppliers. In addition, it should help interested parties who want or need to deal more intensively with the possibilities of application, to think and work through the material in a structured manner.

In addition to conveying the necessary knowledge about this technology, the reader is asked to deal with their own processes. Only if you know your own process structures, can you possibly formulate requirements for third parties, implement these requirements technically.

In recent years, hardly any medium has not dealt with the topic of "blockchain". This emerging technology has left the world of nerds and early adapters and is now on its way to conquer the world. This is mainly achieved by the digital currencies (cryptocurrencies) based on blockchain technology[1] such as Bitcoin, Ethereum or, more recently, Cardano and others.

Enthusiasts therefore also claim that this technology will continue to expand to all other industries and conquer the world in a storm. Critics, on the other hand, see this

---

[1] Cryptocurrencies are so named because encryption and other cryptographic elements play a significant role in avoiding that money can be spent multiple times.

© The Author(s), under exclusive license to Springer-Verlag GmbH, DE, part of Springer Nature 2022
K. Adam, *Blockchain Technology for Business Processes*,
https://doi.org/10.1007/978-3-662-65818-5_1

technology only as a niche technology because it has not yet proven its mass suitability. In order to be able to make an assessment and evaluation yourself, the basic concepts of the technology will be explained in this chapter. This will include a description of the different types of blockchain that have (so far) been established for what purpose and what technical concepts are behind these types.

"Stories of revolutions are stories of the unexpected, almost the impossible, that then happens," write Patel and Moore in their book "A History of a World in Seven Cheap Things."[2] The blockchain technology is often described as a revolutionary and disruptive extension of the Internet. Revolutionary because it makes the classical intermediaries, which we today have in many processes, appear superfluous. Disruptive potential arises from the elimination of intermediaries, as now completely different business processes are necessary. Processes are carried out directly (so-called peer-to-peer processes) and without detours via any intermediaries. Therefore, it is to be expected that these processes can be designed more lean, effective and efficient and thus make traditional approaches superfluous. Just this prospect must shake everyone up because the existing is in danger of becoming superfluous. It is important to recognize which processes are worth being represented on a blockchain. Other processes, however, do not need to be rethought because they, as they are designed, make the most sense. However, blockchain solutions enable us to think beyond the horizon, that is, to approach the impossible. You will be able to draw some unexpected conclusions while reading this book—I hope you have as much fun reading and discovering new possibilities as I do every day when I deal with the technology and its facets.

An important note on the use of the term "technology" should be mentioned here:

In colloquial usage, the term "technology" is used both as the "science of technology" and as the "human-made objects/artifacts" synonymously. The difference should be illuminated as follows:

▶ **Definition** The German Association of Engineers (VDI) defines technology as follows in its guideline No. 3780:

"Technology in the sense of this guideline comprises

- the set of purpose-oriented, artificial, material structures (artifacts or systems);
- the set of human actions and facilities in which systems arise, and
- the set of human actions in which systems are used.

Technology assessment thus not only refers to the material systems, but also to the conditions and consequences of their emergence and use."[3]

---

[2] Patil, Raj; Moore, Jason W. (2018): A History of a World in Seven Cheap Things, p. 272.
[3] VDI guideline 3780, p. 66.

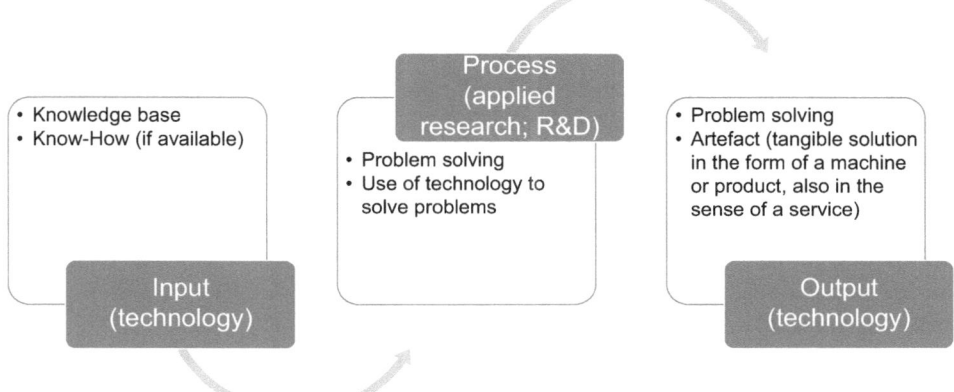

Fig. 1.1 Technology and technique as an application-oriented approach according to Bullinger. (own and supplemented representation)

Technology is therefore either a device, a process or a skill and is also often understood in connection with a trade.

Technology, on the other hand, is the doctrine or science and consists of the two Greek words "techne" for technique and "logos" for logic and/or reason and deals with the possible methods for achieving a defined goal.[4] In addition to the application level, which includes the technical representation, technology also includes the higher-level environment factors and their interaction with artificial physical systems.[5] These include economic, legal, social and social factors, from which, according to Hoffmann, the "decisive extension of the concept of technology to technology" can be derived.[6] Hoffmann refers to the goal-finding process in order to find a solution to multi-layered problems from a variety of sources that goes beyond the mere application-oriented technical level.

This understanding of the difference between technique and technology is supported by Bullinger's application-oriented system approach, which he defines as "[…] the set of all known possible methods for achieving a goal in a delimited application area […]" (cf. Fig. 1.1).[7]

Differentiating the terms used is important in that different levels are targeted with these different terms. It is therefore important to understand the overarching framework

[4] Duden (1963): the etymological dictionary.

[5] Ropohl (1999), pp. 117 ff.: Physical systems are defined as purpose-oriented, artificial physical objects and can be understood as a general term for technical achievements.

[6] Hoffmann (2011), p. 12.

[7] Bullinger (1994), p. 34.

of blockchain technology in order to subsequently illuminate the technique, the application. This differentiation does not take place in the Anglo-American world with the use of the word "technology" and can lead to misunderstandings when simply translating.

That's why it's important to make it clear beforehand which level is being addressed. In the majority of cases, this book will be about the technique, that is, about the question of how and in what context this technique can be used sensibly for a business idea.

If these details and the resulting complexity are understood, the technology will be shrouded in the hype and/or expert knowledge. The more people/decision-makers dare to look behind the scenes, the less they are seduced and blinded by fabulous (and unrealistic) promises that only lead to disappointment in the end. This kind of actionism was well observable on the market in the last years.

My idea with this book is to "de-mystify" this technology and technique and to make it understandable to business-minded people in a company from the corner of pure expert knowledge. It is a balancing act between the necessary terminology with technical details paired with the transfer into business models and ways of thinking. In my opinion, this is necessary because, in particular, in SMEs, the existing research budget is too low and the existing IT department is too much involved in the daily work. Here, the management has to take the lead and build up the necessary knowledge in order to be able to make strategic decisions. The decision may also be against the use of this technology and technology—only: Before this decision is made, one must know what one rejects.

Let us therefore start!

1.1 What is "the" Blockchain Technology?

First of all, there is not "the" blockchain, but the doctrine of the blockchain. This is meant to be the extended approach from the metaperspective and allows the consideration with regard to the interactions between environmental factors (economic, legal, social and social factors) and artificial physical systems (the various blockchain techniques, the questions of "on-chain" and "off-chain" storage [8] etc.).

When referring to a blockchain, this is typically done in the context of cryptocurrencies like Bitcoin. However, the blockchain technology and its field of application are much greater than that they are "only" used as the backbone of digital currencies. The tech community and many other participants are very busy finding other innovative ways to use this technology. A prominent example is the banking industry.

---

[8] "Off-chain" refers to storage processes that are not stored in the blockchain, but in another database.

You can approach this question quite pragmatically by looking at the different types and the associated design options (see also Sect. 1.5) in order to describe the relationships between them.

The blockchain technology, and extended the so-called "distributed ledger technology" (DLT), is based on decentralized data storage and management functions.

▶ **Definition** A blockchain is a distributed database that stores a chain of digital data blocks additively. A blockchain can also be referred to as a distributed ledger (distributed directories), but not every distributed directory that has a decentralized data structure has to be a blockchain. Distributed directories, though, do not have to be chained together like a blockchain.

However, if the distributed data structure is in the form of a chain in which transactions are grouped into blocks and chained together with so-called hashes, then it is a blockchain.

The Federal Office for Information Security (BSI) has represented this multi-layeredness in a graphic (cf. Fig. 1.2).[9]

At its core, the technology is based on a network structure and consensus mechanisms within the network structure (cf. Chap. 2), cryptography (cf. Sect. 3.1), the data structure on the blockchain in general. In interaction with the infrastructure surrounding the core, the questions of network access arise as part of the network structure (cf. Sect. 1.5). This is such a fundamental decision because it determines who is allowed to participate in the network.

In this context of participation and involvement, answers must be found as to which interfaces the participants can be integrated into. In addition, it must be considered which role is assigned to which participant. The rights and obligations arise from the role. This requirement is supported by cryptographic (additional) functions.

This figure refers to the (necessary) interaction with an infrastructure surrounding the core and makes the dimensions visible at the same time. Therefore, one can speak of blockchain technology in this context and at the same time gain an understanding of the great potential of this technology.

In order to exploit this potential, both levels (core and infrastructure) must be able to react flexibly to the requirements placed on them. Blockchain technology as a whole must continue to develop. A brief overview shows which phases this technology has gone through in the past decade since its market launch.

---

[9] Berghoff et al. (2019), p. 11.

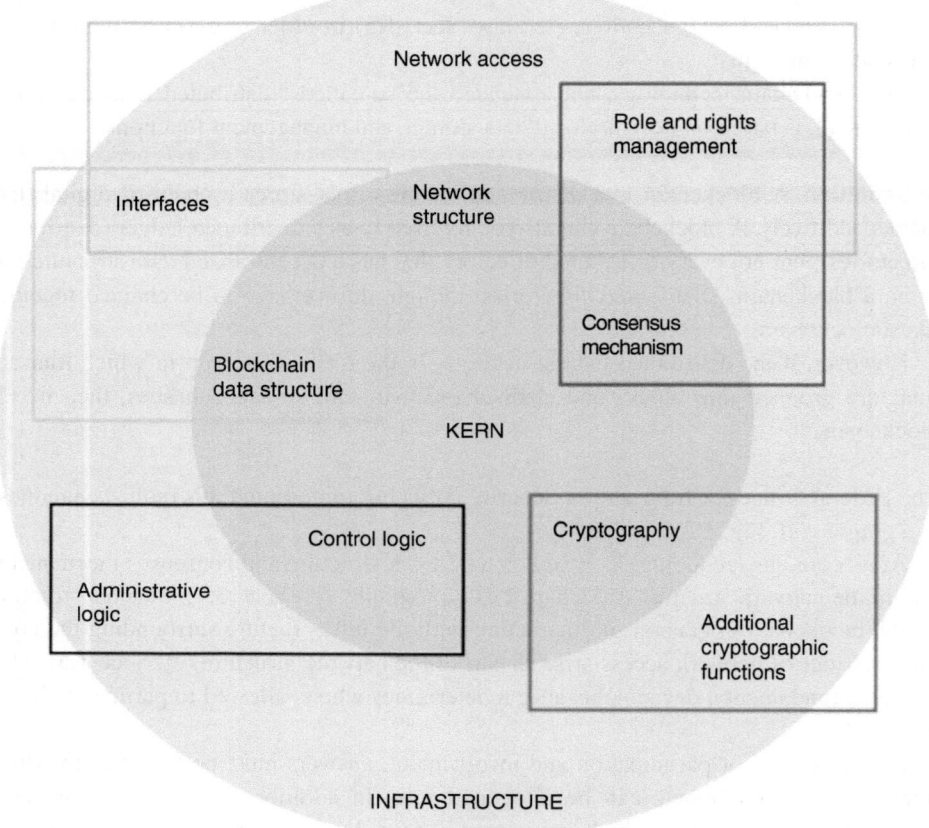

Fig. 1.2 Blockchain-Schichtenmodell. (Aus BSI 2019, Blockchain sicher gestalten; mit freundlicher Genehmigung von @BSI, All Rights Reserved)

This is how one speaks of

- Blockchain technology 1.0 when the use of the technology is primarily for pure financial transactions, as is common with the Bitcoin application.
- Blockchain technology 2.0 deals with more than "just" the transaction of payments. With the extension, assets (at least theoretically) can be transferred between two participants without any intermediary, such as a notary, using so-called smart contracts (more on this in Sect. 3.2).
- Blockchain technology 3.0 not only deals with the previous concepts of its predecessors, but also strives to overcome the limitations that are visible under blockchain 1.0 and 2.0. This includes, in addition to a transaction rate appropriate to the requirements of the economy, questions of energy consumption, block size and scalability.

The expansion of decentralized data storage in a network environment in which the network makes the decisions is the basic basis of this technology. The management functions act inwardly, into the technology, as well as outwardly into the infrastructure. This technology is in the making, not in existence. It sees itself as "open source", i.e. existing copyrights belong to the public, the network. And this network (e.g. via GitHub) also works on the further development of this technology. There is no doubt, however, that this technology inspires and enables economically shaped business models in order to develop innovative products and services with the network structure.

In addition, to shed light on the terminology with another facet, it should be allowed here to point out that, with the progressing digitalization (accompanying the progressing globalization), the so-called platform economy is progressing. Platforms are networks, see e.g. Amazon. Amazon has been offering much more than just books for a long time now, rather the customer can buy everything from A to Z on this platform. With the help of external providers, Amazon systematically increases the range of products on offer. Meanwhile, Amazon is used as a search engine, i.e. customers who are looking for something prefer to use Amazon for their research.

Marketing experts explain that, although we all believe that we know how much influence Amazon and other digital companies have on our everyday lives, we still systematically underestimate their power.[10]

And for this reason, the determination of a new network orchestration is necessary. Blockchain technology can contribute to the fact that the market power of individual companies is used better in favor of a distributed network in the broader context, and this technology will contribute to increased transparency.

1.2   What now is Blockchain Technology?

The blockchain technology first describes a database, but which is distributed on the respective participating nodes or computers and not centrally on one computer. Within this database, blocks filled with data are securely chained together. This chaining is strictly additive, i.e. always a block is attached to the previous one. It is not possible to squeeze a block, a transaction retrospectively into the existing chain of data blocks. The data or transactions in such a block can be transfers, orders, orders, confirmations of authenticity or other certificates and claims.

Put very simply, you can imagine a block as an Excel table to understand the idea. In the case of cryptocurrencies such as Bitcoin, this Excel table has three system-required

---

[10] https://www.searchenginewatch.com/2019/08/01/amazon-google-market-share/, Steve Kraus: „Many people guess Amazon's market share at around 40–50% – but that's how they perform in their worst categories, like clothing and furniture"; accessed on 28.08.2019.

columns: "User A", "User B" and "Amount". If user A now transfers a certain amount of cryptocurrency to user B, this happens directly from computer to computer.

In the Excel table, this process is entered anonymously, i.e. neither user A as sender nor user B as receiver are visible by their names. To actually transfer a digital unit of money, you need a (in this example) Bitcoin address as a public key and a private key. The Bitcoin address can be generated by a wallet provider or generated randomly.[11] Public keys are calculated from private keys using elliptic curves (see Sect. 3.1). The Bitcoin address as a public key is often compared to an email address that everyone knows. But only the one who knows the password can open the mailbox itself. In this example, the password corresponds to the private key. User A therefore uses the private key to sign a transaction for an amount X to a recipient and send it to the network. The network checks whether A has the amount to be sent and then confirms the correctness of the transaction if it has the corresponding credit.

When creating a new email address, you always have to come up with a password. No new email address is generated without the password. Generating the public email address is easy, but with today's computing power it is not possible to calculate the private password from this public address. Blockchain technology also makes use of this property.

The table is updated in blocks and is located on all computers in the network. This means that all other participants in the network automatically keep a book of the transactions made and confirm at the same time that these transactions have actually taken place. The account balances are publicly accessible.

Owners of digital currencies "only" own them in the form of records of inflows and outflows. There are neither physical nor digital units of digital currency.

Currently, many transactions are paper-based or partially digitized. Our entire economy is still based on paper. So we keep our contracts just as much on paper as we vote by paper ballot or manage our bank accounts. Without a paper receipt, there is a lack of trust. This is partly because digital media can be copied and reproduced. Digital media are dynamic in contrast to our known static paper world.

Nick Szabo already states in 1997 that "digital media can perform calculations, operate machines directly, and perform some types of reasoning much more efficiently than humans" (see Szabo 1997 in Formalizing and Securing Relationships on Public Networks). Our current trust in the static, because paper-related world is based on the fact that the contracts and laws surrounding us can be interpreted and enforced by authorities/institutions. But this "paper madness" costs a lot of money, is error-prone and very slow.

---

[11] Bitadress.org for the random generation of a Bitcoin address and the receipt of the private and public key, accessed on 30.08.2019.

For the jump into digitalization to work, trust must exist that the digital world can offer at least as good protection for property, contract content and -loyalty, etc., as we know it from the analog world. For this to succeed, it requires traceability combined with the knowledge that no such data can be manipulated. This claim opens the doors to blockchain technology, because this technology has the potential to close the gap from the real world into the digital world. Data stored on a blockchain are stored on a variety of decentralized networked computers, all of which have the same complex data chains/ information. The attempt to change or even delete a single data block will fail due to this data chain structure. The node/computer points all have the same information and can therefore also control each other. A changed entry in a block leads to a false chain. Since the other participating nodes cannot trace this change, they cannot confirm the value, this new chain remains meaningless—no further transaction is attached to it, because the validation is missing. The system is based on the sufficient confirmation of all participants.

Blockchains can be used both in the inter- and intranet. Their fascination is derived from the versatility of their possible uses. Even if critics—rightly—argue that the breakthrough has not yet been made, the potential is undisputed. It is the challenge to determine the right application case in your own company.

The blockchain industry can already offer modular solutions that the user as a customer has to evaluate for their suitability. In the end, however, every solution must have a value added. For this, extended knowledge is necessary in order to be able to make decisions in the right context.

1.2.1   Use of Blockchain Technology

Usually, this technique is associated with the use of digital currency. But as already explained, blockchain is so much more. However, many projects that were introduced during the peak of the hype have either completely disappeared from the market or continue to exist under completely different conditions today. What is also visible in commercial software development, namely that products are invented that have no market, is also not different for the blockchain world. Using blockchain technology just to present a blockchain-based solution is not sustainable. However, the requirements of an increasingly complex and divisional world must still be captured and classified. In addition, the technology has many facets that allow for modular use.

Dave Snowdon and Mary E. Boone develop a framework (Snowdon and Boone 2007) because they find that despite good training, leaders are not always able to deliver the desired results in situations that demand almost simultaneous decision-making. Perhaps the time component is not the decisive factor in the question of whether blockchain technology should be used in your company. However, both the technical, organizational and economic aspects must be understood in the question, and this is complex. Therefore, it is worth considering the Cynefin Framework by Snowdon and Boone. After a short

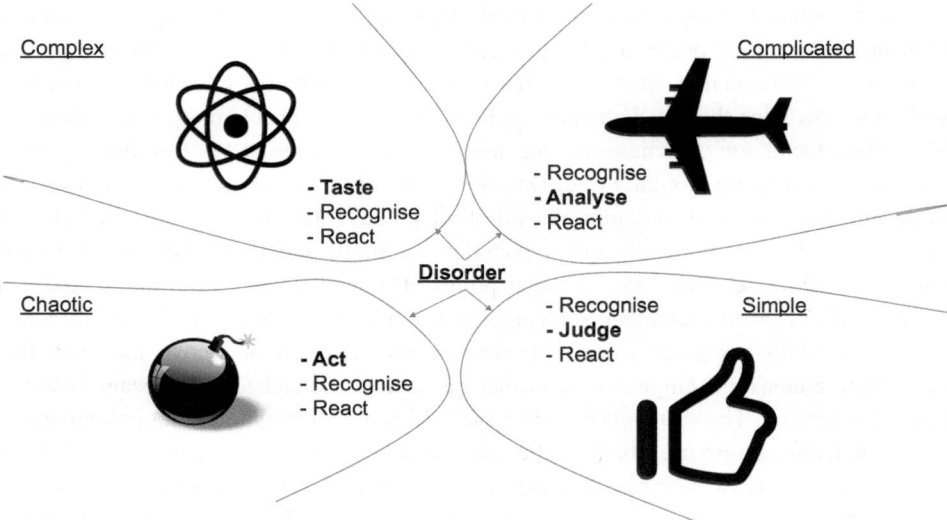

Fig. 1.3 Cynefin Framework

description of the domain properties in this model, this is applied to the blockchain world (see Fig. 1.3).

Snowden and Boone have identified five domains, each with different properties: simple ("simple"), chaotic ("chaotic"), complicated ("complicated"), complex ("complex") and Disorder in the sense of "not knowing, lawlessness".

If you transfer these properties to the blockchain technology and its possible applications, the following picture results:

1.2.2   Simple (Best Practice)

With simple problems, cause-and-effect relationships are obvious and clear. There are solutions that already exist within the company that can be used as templates. If you transfer this approach to the blockchain environment, the result/finding will be that implementing such a solution is not worth it in this case, as it is too complex itself and the advantage for the size of the problem is too small. It is not worth thinking about such processes and the underlying problems using a blockchain solution.

1.2.2.1 Chaotic (Primacy of Quick Action)

Chaotic problems and processes require immediate action. You are forced to act on several crisis fronts at the same time. With such problems and processes, the priority is to avoid further damage and to lead the company out of the crisis mode. A democratic leadership style is of no help here, but rather the authoritarian one, in order to not only make decisions, but also to implement them. This does not apply to the use of a blockchain solution. These intervene very deeply in existing processes and should under no circumstances be used as a quick solution to acute emergencies. (The process steps that have to be gone through in order to come to a sustainable application are shown in Chap. 4).

1.2.2.2 Complicated (Expert Domain)

Processes are so complicated that they require expert knowledge. With the help of these experts, these processes and problems can usually be controlled well. The cause and effect can be found and eliminated by means of a corresponding analysis. This is time-consuming, but manageable. The construction of the Airbus 380 is often cited as an example, where at first the cable strands were too short. It is highly complicated to assemble such an airplane, but with expert knowledge and the analysis of cause and effect, problems can be solved—as with Airbus.

In this quadrant it makes sense for the first time to think about blockchain-based solutions. Although a blockchain solution cannot prevent short cables, as described in the example above, it can contribute to faster error correction through the inherent transparency as well as to a claim settlement of guarantees. Once cleverly set up and mapped onto a blockchain, processes are manageable and contribute to increased efficiency, as so-called intermediaries are no longer or no longer required to the same extent as before implementation.

1.2.2.3 Complex (Innovation)

In contrast to complicated situations, a complex situation is characterized by unpredictability. Facts must be checked and verified and adjusted by iteration. The outcome cannot be determined in advance. Creativity and innovative spirit are important, as is the courage to learn (others may also call it failure—but even from a failed project one can learn a lot).

This is typical Blockchain terrain because the technology can be used in many different ways. It requires a "sandbox" to test. Even routine processes can become exciting again for review when it comes to automating them adequately to relieve employees from paper-based routine work. However, these processes must be carried out with the same reliability and better if they are to be replaced. In the end, these tested processes end up in the "Complicated" quadrant.

1.2.2.4 Disorder (in the Sense of "Not Knowing" or "Disorder")

This "disturbance" refers to the fact that decision-makers do not know on which squares they relate (should) to a situation. If this happens, a decision is made according to one's own preferences. The authors of the model recommend in this case the "disassembly" of the process/problem into individual segments in order to reflect them over the four given quadrants. This is also recommended when considering using Blockchain technology for business processes.

The Cynefin Model illustrates the complex structures in which companies are increasingly finding themselves. The shelf life of assumptions according to which a company can build and expand itself and its strategies is drastically reduced and requires new ways of working and thinking. No wonder, then, that another buzzword, also often used in connection with Blockchain technology, is:

1.2.3   Agile Working or Scrum

The changes in our environment are also reflected in our working world. The change in corporate and working culture can be supported by agile working methods. Responsibility, mutual respect and independent work characterize this change, which must not be dictated from above. The change to the agile company is achieved by many steps. Interdisciplinary team work is promoted.

When it comes to the question of whether the implementation of a blockchain-based solution for one's own company is worthwhile, interdisciplinary small teams are (or should be) used, which have to organize themselves. The expected outcome must have an holistic product development as a goal. The consequence of this is that these self-organized teams must be equipped with decision-making power in order to be able to deliver usable results. Scrum, originally a term from rugby, is extensively illuminated in the article by Takeuchi and Nonaka (1986) in the Harvard Business Review. As early as 1986, the authors find that the traditional sequential product development approach no longer meets the requirements of speed and flexibility. "Under the rugby approach, the product development process arises from the constant interaction of a hand-picked, interdisciplinary team whose members work together from beginning to end" (Takeuchi and Nonaka 1986). This integrated approach, which expressly allows "try and error", makes agile work possible on the different levels and functions in an increasingly complex world.

Programmers are also trained to analyze a problem in sufficient depth first and to create a corresponding solution before actual programming begins. Developers must (also) learn that a completely thought-out solution at the beginning is not always necessary. Achieving sub-goals in stages, which can be adapted to new circumstances, is more effective in finding solutions in increasingly complex processes than a so-called pure sequential processing of the problem.[12]

In Scrum, various terms are predefined, which are explained briefly below (cf. Fig. 1.4).[13]

In Scrum, the product owner is responsible for the economic success of the product. He/she begins the product development with a clear product vision. In line with the

---

[12] Cohn (2010), p. 8.
[13] Rubin (2014), pp. 48 ff.

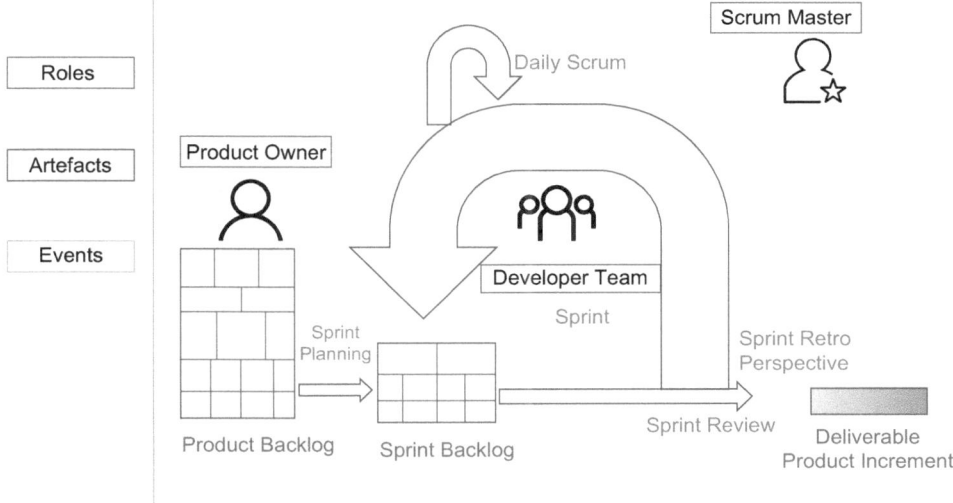

Fig. 1.4 Scrum process

product vision, he/she creates and prioritizes the requirements in the so-called product backlog. He/she involves all relevant parties early on in the definition of the product vision and the product backlog. However, the product owner always remains in control of the product backlog. At the beginning of each development cycle, called a Scrum sprint, the Sprint Planning takes place. During the Sprint Planning, the product owner and the development team agree on which requirements from the product backlog should or must be completed in the sprint. These requirements are transferred to the Sprint Backlog. It is important that the product owner and the development team agree on the Sprint Backlog. This means that the development team itself estimates how much it can do in the next sprint. A sprint has a fixed, always equal length. The development takes place in directly consecutive sprints. The always equal length leads to a good comparability of the sprints, so that in particular the speed of the previous sprints can be quite accurately predicted from the speed of the previous sprints.

Many teams work with two or three weeks as a fixed sprint length. But one or four weeks are also not unusual. During the sprint, the self-organized team works through the sprint backlogs without interruption. This also means that no new or changed requirements are made to the team during the sprint. This is only possible after the end of the sprint in the Sprint Planning of the next sprint.

The Daily Scrum provides for the daily status check and the deployment planning for the day. The Daily Scrum takes place every workday at the same time and lasts a maximum of 15 min. Therefore, it is usually carried out standing. In turn, each team member answers three questions:

1. What have I accomplished since the last Daily Scrum?
2. What is hindering me in my work?
3. What will I have done by the next Daily Scrum?

The Daily Scrum does not include discussions. These can usually take place in smaller groups afterwards.

The Scrum Master ensures that Scrum works in the team. He/she is responsible for ensuring that the Scrum process is adhered to, helps to remove obstacles, protects the team from external disturbances and helps the team with self-organization.

He does not directly solve the team's problems, but helps the team solve problems itself. The Scrum Master is a leader in a modern sense. He/she is not authorized to give orders to the team and follows the principle of "leading by serving".

At the end of each sprint, the sprint review takes place. Here the team presents to the product owner what it has achieved in the sprint. For this it is important that the requirements are broken down in the sprint planning so that there is something to present and see after each short sprint.

Often new requirements arise during the sprint review, which the product owner then enters directly into the product backlog.

While the sprint review is about looking at the result, Scrum teams look at their process in retrospectives. They ask themselves what went particularly well and how these properties can be preserved. But they also ask themselves what went less well and how these aspects can be improved.

This is a continuous improvement process built into agile software development processes, based on the fact that we are constantly learning.

Now the next sprint can begin.

Scrum helps to iteratively manage the changes that occur in complex product development. A potential blockchain implementation in an existing corporate environment with existing IT is highly complex. Scrum will not provide answers directly, but rather encourage the team to ask questions. This makes it possible to discover and fix malfunctions more quickly.

▶   Please also allow the questions that "hurt", because they confront you with
    weaknesses. Even if human nature, especially in complex processes, rather
    simplifies to make decisions. In addition, we have an affinity for "preservation"
    in us. But that means that when we have to make decisions, we tend to "pre-
    serve" against changes. Critical questions help to break this pattern, even if it
    hurts!

Sometimes it is suggested that Scrum is simple because the framework fits on a famous beer coaster. To assume alone that the speed and quality of processes and products increases through agile work is a mistake. It is much more complex. Traditional, sequential processing of projects does not lose importance just because everything has to be agile today. Rather, it is important to recognize when techniques like Scrum can bring projects forward in a meaningful way.

Since Blockchain technology or "Distributed Ledger Technology" is also very complex, the processes have to be iterated again and again in order to take into account changes in a timely manner, the use of Scrum is recommended.

1.3   Characteristics of Blockchain

The concept of blockchain technology is explained in the white paper by Satoshi Nakamoto in 2008, without him using this term explicitly. The cryptographer David Chaum describes in 1982 how he can separate the connection between digital messages, which is generally considered the foundation for realizing anonymous payments.[14] The analogy of cash to digital money lies in anonymity: Just as with cash, it is not possible to trace who the owner of a banknote is (in contrast to credit card payment, with which a bank can very well trace how and where a customer spends his/her money). With eCash, Chaum launched digital money for small transfers in the mid-1990s as one of the first predecessors of Bitcoin.[15]

A little later, in 1991, Haber and Stornetta published their first approaches to the secure linking of data blocks. The authors describe a technique with which they can timestamp digital documents and verify the authenticity of the document in order to establish the authorship of the document.[16]

Shortly afterwards, in 1994, Nick Szabo described smart contracts, but at this time his/her theory of how intelligent contracts could work remained unfulfilled because there was no technology to support programmable agreements and transactions between parties.[17] In 1998, Szabo introduced Bit Gold, a scenario in which he created individual currency units using a work algorithm almost identical to Bitcoin's.[18]

In 1997, Adam Back described how computing power can be used to create value. To do this, he uses the proof-of-work approach to reduce spam e-mails by requiring the sender of an e-mail to provide a certain amount of computing power as a fee. When sending an e-mail, this fee is not noticeable in terms of computing power. However, with mass and spam e-mails, this mechanism can cause undesirable delays.[19] Even though

---

[14] Chaum (1982).

[15] However, the company digicash founded by Chaum went bankrupt at the beginning of the millennium.

[16] Haber/Stornetta (1991), p. 99.

[17] Szabo(1997) [a].

[18] Szabo (1998/2005) [b].

[19] Morabito (2017), p. 10.

this concept never really took off for e-mails, with Bitcoin this proof-of-work is an important prerequisite for the so-called mining.[20]

In 1998, Wei Dai developed his b-money project, which aims to be an "anonymous, distributed electronic payment system". In this way, many services and functions are provided that also modern cryptocurrencies provide.[21]

In 2005, Hal Finney attempted to describe digital money and drew on the work of his predecessors. But even his approach did not reach the market. Finney's variant provides for a central unit to make the money available to the users.

Central services are of high importance in the world of the Internet.

Online services that are still used by the majority today are based on the so-called "client-server model", in which the server acts as a service provider and provides the user with functions and resources and takes on a variety of roles/tasks.[22] Some of these server functions are, for example, responsible for creating marketplaces between buyers and sellers (as in the case of ebay and Uber). Others are responsible for storing and maintaining data stores that are collected by various parties on the Internet (examples include Facebook, Youtube and Wikipedia). Others again serve as authenticated sources for certain goods or services (e.g. PayPal or Spotify). The customer (client) uses the services made available to him. Overall, this is a centrally oriented architecture.

Blockchain networks act as decentralized global networks differently. Via an all-encompassing, cross-protocol software, the computers are networked with each other on an equal footing, and there is no need for a central instance for the maintenance or operation of the blockchain.[23] Since it is not equipped with a central instance, in theory anyone with an Internet connection can access the information stored on a blockchain by simply downloading freely available open-source software.(This is at least the case with public blockchains).

Thus, blockchain networks overcome the limitations described above and enable people to store unassailable data pseudonymously and transparently across national borders.

In addition, blockchains, like other databases, are also expected to reflect the facts correctly. This expected data integrity is ensured by various measures such as regular updates of backups, log-in files and access control to the data stored in the database. Traditional databases are centrally organized, and so in the event of a successful attack, the data is at worst no longer available. The decentralized database structure of a blockchain avoids this problem because the backups are located on all participating nodes. This is referred to as data integrity and "No-Single-Point of Failure".[24]

---

[20] Back (2002).

[21] Wei Dai (1998).

[22] Bryant/O'Hallaron (2016), p. 954.

[23] Kersken (2019), p. 203.

[24] Berghoff et al. (2019), p. 18.

Other features that are attributed to the technology:

- Decentralization: All data is stored on the network's equally privileged computers (peer-to-peer network),
- Trust is created because the different computers involved, known as nodes, cannot be manipulated, as certain procedures ensure that only valid data sets are accepted and data integrity (see above) is maintained. This creates trust in the system.
- Transparency: The data on a blockchain is traceable and transparent through verifiable activity paths.
- Anonymity: At least pseudonymity can be guaranteed, as participants are represented by encrypted numbers and letters, not by a real name.

Thus, a blockchain is essentially a database that stores information chronologically in a constantly growing chain of data blocks and implements it in a decentralized network so that data integrity, trust and security are created for the nodes without the need for central authorities or intermediaries. In its most concrete form, it is a computer code that tells every computer on which it is implemented to store data locally. Thus, it is part of a global network with thousands of other computers and also stores data with the same (conformant) programming code.

1.4 Decentralization

Decentralization and blockchain are often mentioned in the same breath. Therefore, it is worth taking a brief look at decentralization.

At the beginning of the Internet until the early 2000s, Internet services were based on so-called open protocols, which were checked by the Internet community. Organizations, companies and individuals can set up and expand their Internet presence, knowing full well that the rules of the game on the Internet do not change (will not change). In this time, the big Internet companies such as Amazon, Facebook, Google but also companies like Alibaba and Tencent start their triumphal march. With the beginning of the second era of the Internet—around the middle of the first decade of the 2000s, continuing to today—profit-oriented tech companies (in particular Google, Apple, Facebook and Amazon—abbreviated GAFA) develop software and services that go far beyond the possibilities of open protocols. This trend is supported by the almost explosive increase in the worldwide spread of smartphones, as mobile apps make up the majority of Internet usage.[25] Internet users access the network by making use of the service and software offerings of the tech giants. As convenient and easy as this is on the one hand, on the

---

[25]Digital Index 2017/2018, e.g. p. 8; according to Statista, 57 million people in Germany used smartphones in everyday life in 2018; this represents an increase of 3% compared to 2017.

other hand these tech giants dominate the Internet—and they can determine the rules at their discretion (cf. e.g. Google ranking factors). One answer to this centralization and the associated power is the introduction of state regulation—if the Internet could be considered as a hardware-based network (like radio and television stations). The Internet, however, is the ultimate software-based network. This can be re-structured and decentralized by entrepreneurial innovation and market shifts. Software is the coding of human thought and as such has almost unlimited design space!

We have seen the market power of centrally organized platforms like Google, Facebook etc. These platforms do everything to make their services (for us users) as valuable as possible. Since platforms are systems with multi-sided network effects by definition, this added value can be mapped to a certain extent. This allows these providers to give millions of users access to technologies and applications and to do so largely without additional charges. However, the awareness that each user of these platforms "pays" with his/her data is increasingly gaining ground, and with it the need for new, user-friendly solutions. With decentralization, it is possible to form networks that are based on the interests of the Internet community in order to gain acceptance. This method worked well in the very early phase of the Internet, but no longer does so today. The crypto/blockchain world has solved this problem by, among other things, offering economic incentives to developers and other network participants (e.g. in the form of so-called tokens). Blockchain networks use many mechanisms to ensure neutrality even when growth occurs. For example, contracts and transactions are based on so-called "open source codes".[26] In addition, the network participants are equipped with the rights "Voice" and "Exit". This allows the participants to shape the governance, the self-conception of the community. The participants receive a voice both "on chain" (via the protocol) and "off chain" (via the social structures around the protocol). The right of exit allows the participants, for example, to sell their tokens or coins and leave this network or, in the extreme case, to bring about a so-called fork. A new version of the protocol is created here. This is relatively unproblematic with open-source codes, for example, new functionality is added to the open-source code. If enough network participants follow the new development, then there are two versions of a blockchain (as was the case with Ethereum in 2016). In short, it can be said that blockchain-based networks enable their participants to achieve a common goal. In the case of a cryptocurrency, this may be the appreciation of the coin/token and/or the growth of the network itself.

Through this new understanding of decentralized networks, which are determined and shaped by the community, networks arise whose abilities will go far beyond those of the most advanced centralized services. Therefore, it is necessary to plan this into the next generation of business development in order to not jeopardize the previous success of the enterprise. The following example should make this clear:

---

[26] Open-source code is a concept according to which everyone can view, use, further develop and change the source code; see also Gabler Wirtschaftslexikon, Dr. Markus Siepermann.

As Wikipedia threatens to compete with centralized competitors such as Encyclopaedia Britannica and Microsoft's software version Encarta in the 2000s, the makers of the world's dictionaries feel relatively safe: their works are better and more accurately researched. While both Encarta and Encyclopaedia Britannica have editors responsible for collecting and processing information, it is the active network that now offers more articles on more topics, which are maintained and updated by the community.[27]

As a result of the above, it can be concluded that dynamic decentralized systems are superior to static centrally organized systems, as they do not depend on a manageable number of employees and their skills, but rather on the resources of the entire community. Building and fulfilling this community will make future platforms successful. Blockchain-based networks not only benefit from the approach of decentralization, but also from the trust that is created when data is stored and traceable on blockchain applications.

The next section will look at the different types of blockchain and the protocols involved.

1.5   Blockchain Types and Protocols

There is no "the" blockchain, rather it is necessary to distinguish which type of blockchain can be used for which purpose. In order to be able to better classify these differences and the resulting logical consequences, the typical types are described below (see Fig. 1.5).

It all started with the Bitcoin blockchain. This blockchain as a public blockchain can be viewed and downloaded by anyone interested. It is "open source", i.e. free software. The public blockchain is a public decentralized database that is accessible to anyone in the world with a computer and internet access.[28] Within this public database are data of transactions that were carried out by the participants of the network. Through the multitude of users who have the same data stock, a distributed network is thus created. No approval is required to participate in the consensus process, and anyone is allowed to make transactions. The networking of the participants with each other takes place via a "peer-to-peer network". Such networks work without a central server. All participants within the peer-to-peer network (P2P) have the same rights, i.e. each participant is server

---

[27] https://www.forbes.com/2009/03/30/microsoft-encarta-wikipedia-technology-paidcontent.html#73b610862db3, accessed: 07.11.2019.

[28] Luke Parker (o. J.), unter https://magnr.com/blog/technology/private-vs-public-blockchains-bitcoin/, zugegriffen: 15.07.2019.

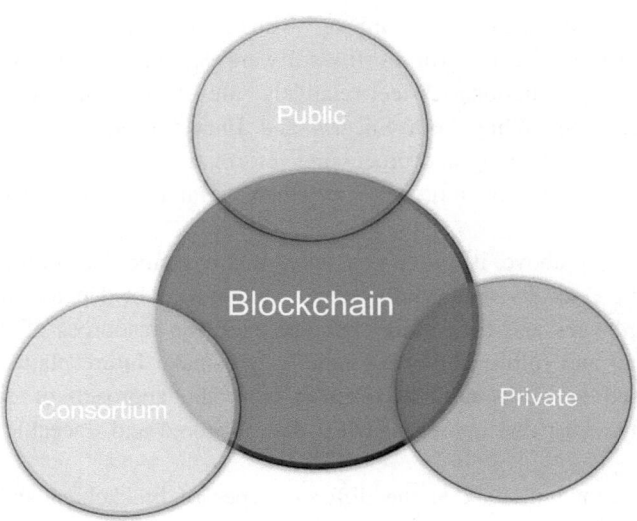

Fig. 1.5 Blockchain-types

and client at the same time, which ensures a high stability of the network.[29] All participants of the network have the right to access the transactions at any time, or to carry out
transactions independently. It is usual to hide the actual identity of all associated participants.[30] This openness ensures the security of the blockchain as well as the ability to
resist hacking attacks or capital controls.[31] The distributed nature of the network, which
verifies the integrity of transactions and the associated account balances, makes a successful attack mathematically very difficult to virtually impossible.

Many companies find this public accessibility to be too deep an intrusion and insight
into their own processes and procedures. For this reason, there is the private blockchain
approach. Here it is clearly defined who has which access to this blockchain in what
form. This can range from simple read rights to administrative rights. Purists criticize
this and also the consortium solution, [32] that the original idea of granting everyone

---

[29] Mahlmann and Schindelhauer (2007), S. 6.

[30] Vitalik Buterin (2015) unter https://blog.ethereum.org/2015/08/07/on-public-and-private-block-
chains/, zugegriffen: 15.07.2019.

[31] Korolov (2016) under http://www.csoonline.com/article/3050557/security/is-the-blockchain-
good-for-security.html , accessed: 15.07.2019.

[32] When companies within an industry join forces to jointly manage data and transactions according to pre-defined rules via a blockchain for a pre-defined application scenario, this is referred to as
a consortium approach.

access to these technical possibilities is in vain. On a private blockchain, the initiator determines who can perform which actions based on which data. Private as well as consortium blockchains use the elements from the public blockchain, but access to the blockchain is centrally regulated.[33]

In addition to the limited access rights and the centralized approach, however, private blockchains can be adapted to new requirements more quickly, since the administrative instance consists only of a few decision-makers. The question of how the nodes in a private blockchain validate transactions and whether or not they need to be rewarded for this also does not arise. As the supreme authority, the initiator can determine how the participating nodes validate transactions and whether or not they need to be rewarded for this.

Another way of categorizing is the approach of the access permission ("permissionless" or "permissioned"): This categorization also differentiates between private and public blockchains. Access without conditions is therefore "public & permissionless", i.e. anyone can participate in these models and validate processes in the blockchain (e.g. Bitcoin). An internal blockchain that only allows a selected group of participants access to the technology would therefore be "private & permissioned". Hybrid solutions combine both areas in such a way that on the one hand a private blockchain is built, but on the other hand allows access "permissionless",[34] or on the other hand a blockchain that is publicly accessible, but requires the acceptance of certain governance rules.[35]

Fig. 1.6 summarizes the approach.

In a centrally organized database, you usually do not need a blockchain approach, unless you want to share the content of the database and need proof that the shared data is stored securely. Depending on the type of blockchain selected, access to the data and validation of individual transactions can be designed. In a decentralized, distributed network, each participant is connected to every member of the network without the need for a higher, regulating authority. Rather, the computers of the participants form the relevant nodes and thus the distribution system (cf. Table 1.1).[36]

The challenge lies in deciding for one's own application how access to the blockchain is to be created and designed. The consensus models explained in the following Chap. 2 show the diversity with which decisions on transactions and validity can be made within a blockchain network.

---

[33] See, for example, Drescher (2017), p. 215.

[34] An example of this is Hyperledger Sawthooth.

[35] In this case, for example, Ripple can be mentioned as an open-source protocol for a payment network.

[36] Drescher (2017), p. 15.

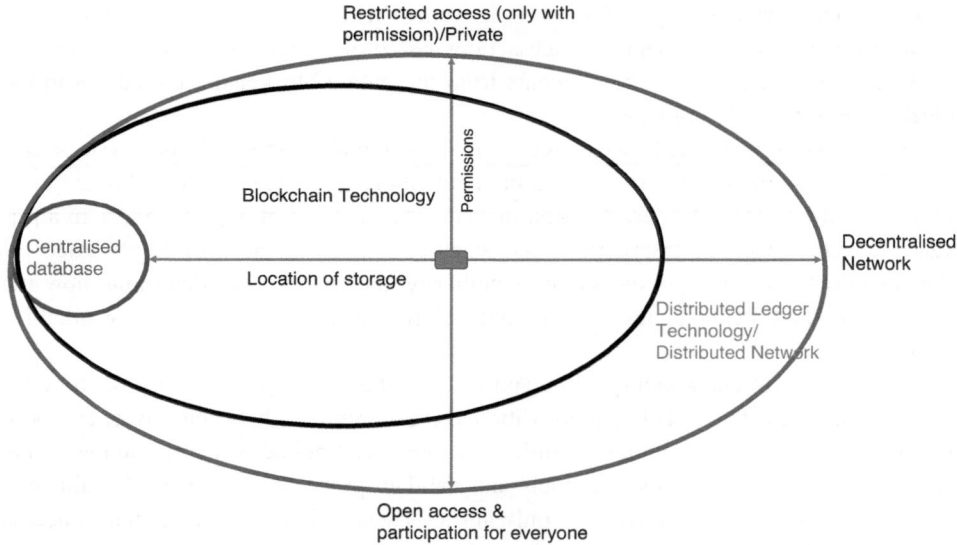

Fig. 1.6 Blockchain categorization

Table 1.1 Characteristics of access possibilities

	"Permissionless" (public)	"Permissioned" (private)
How do you access the network?	Open access for everyone	Authorized access
Who are the validators/testers?	Anonymous, fully decentralized testers	Pre-selected, trusted testers
How are laws and regulations incorporated?	Aims to create censorship-resistant, anonymous transactions—not always compliant with applicable legal framework	Enables the "simpler" implementation of regulations and laws (e.g. KYC[a], AML)
Areas of application?	Open-access applications	Enterprise-owned systems

[a]The acronyms KYC and AML stand for "Know your Customer" and "Anti Money Laundering". KYC refers to customer verification, i.e. the identification of the new customer. AML stands for the Anti Money Laundering Directive. This directive describes the procedure, laws and regulations to be followed in order not to engage in money laundering

1.6   Summary

The distinction between whether one speaks of the technique or the technology in the broader sense helps to distinguish. Although a technique is not detached, but on the technological level the interaction takes place with the extended environment, which it is necessary to understand and to design.

When using the "pure" blockchain technique, it must be checked in which context this solution is aimed at. Iterations help to make the goal more granular. The mentioned models and techniques help to clarify which mental effort is necessary to create one's own reference framework.

The decentralization goes beyond the hitherto conventional extent in terms of blockchain: in fact, in a blockchain network everyone is connected to everyone—in contrast to the previously known. The associated advantages, e.g. in terms of security, on the other hand, call for a higher degree of complexity. Agile methods such as Scrum help to cope with this complexity.

Anyone considering creating their own blockchain solution will quickly find themselves asking about access and validation of individual transactions. The different models have different advantages and disadvantages. If transparency and verifiability are to be in the foreground in a public blockchain, confidentiality of the data to be stored is tendentially more in the foreground in a private solution. Therefore, every team that deals with questions of this kind must also consider what goal is to be achieved.

References

Back A (2002) Hashcash – A denial of service counter measure. http://www.hashcash.org/papers/hashcash.pdf.. Accessed: 15. July 2019

Berghoff C, Gebhardt U, Lochter M, Maßberg S et al (2019) Blockchain sicher gestalten, Konzepte, Anforderungen, Bewertungen (Bundesamt für Sicherheit in der Informationstechni, Hrsg). Bundesamt für Sicherheit in der Informationstechnik, Bonn

Bryant RE, O'Hallaron DR (2016) Computer systems, a programmer's perspective, 3. Aufl. Person Education Limited, Harlow/Essex

Bullinger H-J (1994) Einführung in das Technologiemanagement – Modelle, Methoden, Praxisbeispiele. Teubner, Stuttgart

Bundesamt für Sicherheit in der Informationstechnik (2019) Blockchain sicher gestalten, Konzepte, Anforderungen, Bewertungen. Appel & Klinger Druck und Medien GmbH, Schneckenlohe

Buterin V (2015) On public and private blockchains (Unter Mitarbeit von Vitalik Buterin, Hrsg). Ethereum Blog. https://blog.ethereum.org/2015/08/07/on-public-and-private-blockchains/.. Accessed: 15. July 2019

Buterin V (2016) A proof of stake design philosophy. https://medium.com/@VitalikButerin/a-proof-of-stake-design-philosophy-506585978d51. Accessed: 15. July 2019

Buterin V (o. J.) Ethereum white paper: A next generation smart contract & decentralized application platform. http://blockchainlab.com/pdf/Ethereum_white_paper-a_next_generation_smart_contract_and_decentralized_application_platform-vitalik-buterin.pdf. Accessed: 15. July 2019

Chaum DL (1982) Blind signatures for untraceable payments. University of California, California

Cohn M (2010) Agile Softwareentwicklung: mit Scrum zum Erfolg. Addison-Wesley, München

D-21 Digital-Index (2017/2018) Jährliches Lagebild zur digitalen Gesellschaft. Initiative D-21 e. V, Berlin

Dai W (1998) b-money. https://nakamotoinstitute.org/static/docs/b-money.txt.. Accessed: 15. Juli 2019

Drescher D (2017) Blockchain basics: a non-technical introduction in 25 steps. Apress, New York

Duden Band 7 (1963) Das Herkunftswörterbuch der deutschen Sprache. In: Drosdowski G et al (Hrsg) Duden Etimologie. Bibliografisches Institut AG/Dudenverlag, Mannheim

Forbes (o. J.). https://www.forbes.com/2009/03/30/microsoft-encarta-wikipedia-technology-paid-content.html#73b610862db3.. Accessed: 7. Nov. 2019

Haber S, Stornetta WS (1991) How to time-stamp a digital document. J Cryptol 3(2):99–111

Hoffmann O (2011) Innovation neu denken – Histozentrierte Analyse der Innovationsmechanismen der Uhrenindustrie. Springer Gabler, Berlin

Kersken S (2019) IT-Handbuch für Fachinformatiker, Der Ausbildungsbegleiter, 9., erw. Aufl. Rheinwerk, Bonn

Korolov M (2016) Is the blockchain good for security? Unter Mitarbeit von Maria Korolov (Hrsg. v. CSO). http://www.csoonline.com/article/3050557/security/is-the-blockchain-good-for-security.html.. Accessed: 15. Juli 2019

Mahlmann P, Schindelhauer C (2007) Peer-to-Peer-Netzwerke. Algorithmen und Methoden. Springer (eXamen.press), Berlin

Morabito V (2017) Business innovation through blockchain, the B^3 perspective. Springer International, Cham

Nakamoto S (2008) Bitcoin: a peer-to-peer electronic cash system. Satoshi Nakamoto Institute. https://nakamotoinstitute.org/literature/bitcoin/. Accessed: 14. Apr. 2019

Parker L (o. J.) Private versus public blockchains: is there room for both to prevail? (Hrsg. v. Magnr). https://magnr.com/blog/technology/private-vs-public-blockchains-bitcoin/.. Accessed: 15. Juli 2019

Patel R, Moore JW (2018) Entwertung, Eine Geschichte der Welt in sieben billigen Dingen. Rowohlt, Berlin

Ropohl G (1999) Allgemeine Technologie – Eine Systemtheorie der Technik, 2. Aufl. Carl Hanser, München

Rubin KS (2014) Essential Scrum, Umfassendes Scrum-Wissen aus der Praxis. mitp, Heidelberg

Search-engine Watch (o. J.). https://www.searchenginewatch.com/2019/08/01/amazon-google-market-share/. Accessed: 1. Aug. 2019

Snowden D, Boone ME (2007) A leader's framework for decision making. Harv Bus Rev 85(11):68–76, 149

Szabo N (1997) Formalizing and securing relationships on public networks, pdf. https://nakamotoinstitute.org/formalizing-securing-relationships/.. Accessed: 19. Apr. 2019

Szabo N (1998) Bit gold. https://nakamotoinstitute.org/bit-gold/.. Accessed: 19. Apr. 2019

Szabo N (2005) Bit gold. Satoshi Nakamoto Institut. https://nakamotoinstitute.org/bit-gold/.. Accessed: 19. Apr. 2019

Takeuchi H, Nonaka I (1986) The new new product development game. Harv Bus Rev 1:137–146

Consensus Models

2

Abstract

The choice of the appropriate consensus model determines the design of the block-chain to be used, with some models more suitable for the use of private or consortium approaches. It is determined how the cooperation is designed. The roles that have to fulfill different tasks are assigned.

The different types of blockchain models have many similarities and work similarly, but differ in the way in which the matching of the transaction execution is to be carried out. It is a question of which of the transactions actually made are legitimate and how these transactions are to be attached to the existing blockchain. Various consensus mechanisms are listed below, it should already be mentioned here that this list cannot be exhaustive, because new mechanisms are always being checked and tested.

A consensus mechanism is initially a protocol that ensures that all nodes[1] of the corresponding blockchain network are synchronized with each other and on the basis of which it is decided which of the transactions made in the network are legitimate and thus to be attached to the existing blockchain. This agreement is important for the different types of blockchain in order to enable smooth operation. Every participant in a block-chain network can submit any possible transaction. However, it must be ensured that it is an executable, reliable transaction and not a fake.

---

[1] In blockchains we then speak of nodes when it is a physical network device that can receive, send, create and transmit messages, i.e. any internet-enabled device that can provide the aforementioned functions.

© The Author(s), under exclusive license to Springer-Verlag GmbH, DE, part of Springer Nature 2022

K. Adam, *Blockchain Technology for Business Processes*,

https://doi.org/10.1007/978-3-662-65818-5_2

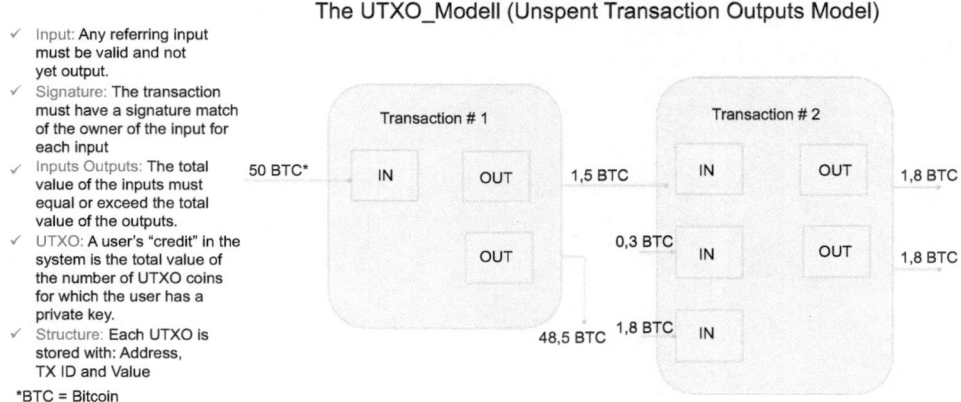

The UTXO_Modell (Unspent Transaction Outputs Model)

✓ Input: Any referring input must be valid and not yet output.
✓ Signature: The transaction must have a signature match of the owner of the input for each input
✓ Inputs Outputs: The total value of the inputs must equal or exceed the total value of the outputs.
✓ UTXO: A user's "credit" in the system is the total value of the number of UTXO coins for which the user has a private key.
✓ Structure: Each UTXO is stored with: Address, TX ID and Value

*BTC = Bitcoin

Fig. 2.1 UTXO model

Example: Let's say person B has three units of a digital currency and wants to send one of those units to person C. The network therefore has to check whether B actually has the corresponding currency units in order to send one of them to C. This has to be logged in order to ensure subsequent transactions in the network that B now only has two units and therefore cannot spend any more until he or she receives new units. C, on the other hand, has one more unit than before and can therefore also dispose of it. If B now comes up with the idea of transferring four currency units to person E, B can communicate this verbally (or in other ways), but the network simply will not allow this transfer because B does not have these (four) units. This would make the last transaction invalid, it would not be carried out by the network, and this transaction would not be added to the blockchain. This is called the UTXO as "unspent transaction output" because the blockchain, which stores all transactions without gaps, can prove which currency transactions have flowed from which user to which user or to which address (cf. Fig. 2.1).

General procedure for a transaction (including a transfer) on a blockchain:[2]

1. *Transaction:* In this first step, the sender creates a transaction that contains information about the recipient's public address, the value of the transaction, and a cryptographic digital signature that verifies the validity and credibility of the transaction.
2. All transactions within a time window are sent to the participating nodes of the blockchain. (If the number of transactions to be executed exceeds the capacity of the network at that moment, the miners as validators can decide which transaction to store when in which block. If a user pays transaction fees, his/her chances of having his/her transaction validated and executed quickly increase.)
3. *Transaction Authentication:* When the nodes in the network receive the transaction (s), they first validate the message by decrypting the digital signature.

---

[2] Froystad/Holm (2015), p. 11.

4. *Block Creation:* Each node collects pending transactions to validate them and aggregates them into a block. Then, using the consensus algorithm "Proof of Work", the correct solution for the block is calculated.
5. *Block validation:* The node that found the correct solution for the corresponding block shares its solution with the network and receives the reward for the computational effort expended.
6. The transactions in the block are accepted by the network if all the transactions contained in it are valid.
7. *Block chaining:* If all the transactions in a block are approved, the new block is "chained" to the current block chain, which results in the new state of the block being transferred to the rest of the network. The next block can be validated. In addition to the transactions, it contains the hash value of the last validated block.

There are a variety of ways to achieve this consensus, and at least as many ways to criticize and reject it. Therefore, it is worth looking at the different models with their advantages and disadvantages.

2.1   Proof of Work (PoW)

The best-known mechanism is that of the Bitcoin blockchain, which has also found numerous imitators among other cryptocurrencies. The Proof-of-Work (PoW) process[3] is also referred to as mining, because with it the bitcoins are "mined". The nodes act as miners who solve complex mathematical puzzles whose difficulty increases exponentially over time and with the computing power used.[4] The first person to solve the puzzle is allowed to create a block and receives bitcoins as a reward for creating a block.[5] The task is to decrypt data blocks. This is costly and time-consuming, and it is a so-called random process with a low probability. As a result, on average, a lot of "trial and error" is necessary to decrypt a data block. The verification of the correctness of the found solution, on the other hand, is simple and allows the other participants of the network to confirm the value found once.[6,7]

The newly introduced term "hash" or hash value requires a closer explanation, as does an important term for mining, "nonce".

---

[3] Nakamoto, S. 3.

[4] Fertig/Schütz, S. 96, 97.

[5] In 2022, each miner still receives 6,25 bitcoins for successfully creating a valid block. Even if in the summer of 2022 the Bitcoin price, similar to many traditional stock market values, has lost a lot of value (to almost 18,000 euros), a successful mining still corresponds to an equivalent value of approx. 112,500 euros - and that approx. every 10 minutes.

[6] Nakamoto, S. 3.

[7] Blockchain Demo (2019) vgl. https://anders.com/blockchain/hash.html—on this website you can try how long it takes to get a data content and the corresponding hash value that starts with at least four zeros, as required by the Bitcoin Blockchain; accessed: 05.08.2019.

A hash value is the result of a hash function in the form of a previously defined character set. There is a distinction made between open and closed methods. With the open methods, it is possible that different input values are assigned to the same hash value. With the closed methods, to which cryptographic hash functions belong, which are used in all blockchains, this function is collision-resistant. This means that different input values (texts, data) must not be assigned to the same hash value. In addition, cryptographic hash functions are so-called one-way functions. This means that it is not possible to deduce the input value from the hash value.

For example, the content of a book can be reduced to a hash value with a corresponding length by means of a hash function.[8] In connection with hashed information on a blockchain, not only currency transactions can be stored, but also texts, files. And if a letter is changed in these texts, the hash value changes. The following example makes this clear:

Example

For the sentence *"Let's go to the Italian tonight"* the following hash value is output with SHA256: eca7d8cec46f7bb6f4bf99e8bbb11662f1fe76e62479f6fc2cc89f2bdf6112e1.

If we change the sentence to *"Let us go to the Italian tonight"*, we get the following hash value: a827eeb6ce82ccad50d31c31d18385b34e1b99f801871e4e34eaf0ab-d6b41cf7. ◀

It can be seen at first glance that not only the first digit has changed in the hash value, but that it is a completely new, different hash value. Only one letter has been added to make the "Let`s s" into a "Let us **t**".[9] So the hash not only identifies this text, but really the exact version of this text. Every tiny change creates a new hash that serves as an ID for this new version.

And in the example chosen here, it is not possible to infer the input text "Let's go to the Italian tonight" from "eca7d8cec46f7bb6f4bf99e8bbb11662f1fe76e62479f6fc2c-c89f2bdf6112e1" (one-way function).

▶　　The conversion of input values of arbitrary length into output values of fixed length is called hashing. The output value, or hash value, does not allow the reverse calculation of the input value.

In addition, it should be pointed out that due to the distributed, simultaneous nature of this process, it is sometimes possible that more than one node finds a winning hash at the same time. Each winning node adds its own proposed block to the blockchain and sends

---

[8] If we use SHA 256, all information in the data field is reduced to a character length of 64 characters (SHA 256 = 256 Bit = 32 Byte = 64 Characters).

[9] Hashgenerator (2019)You can test this yourself at: https://passwordsgenerator.net/sha256-hash-generator/. Accessed: 15.11.2019.

it over the peer-to-peer network. (Peers are participating computers in this case). In such cases, there is a temporary branch in the block chain, where some nodes add blocks to one branch, while other nodes add blocks to other branches, based on which "winning node" [10] is closest to them. However, when additional blocks are added to these forks, the protocol ensures that the branch with the maximum proof of work (i.e. the longest branch) is included in the blockchain and others are discarded. This leads to a match between all nodes in terms of the state of the blockchain.

In addition to hashing, the term "nonce" should also be mentioned here. Nonce stands for "number that can only be used once", and this value must be found by mining. This closes the circle: The search for the matching nonce for the next block takes time, which in turn means that not all transactions from the network can be processed at the same time, and the described queues of pending transactions arise. [11]

Blockchain variants that do not provide mining in their consensus mechanisms do not require the "nonce" field within their blocks.

The Bitcoin PoW consensus algorithm works well in an open environment where an arbitrary number of nodes can participate in and start mining the network. Currently, there are 12959 nodes responsible for processing transactions.[12] No knowledge or authentication of participants is required, making such a consensus model extremely scalable in terms of supporting thousands of nodes. Thus, this is a completely decentralized network that works.

However, the Bitcoin PoW consensus is vulnerable to "51% attacks", in which a mining pool that is able to control 51% of the mining power (i.e. hash rate) can write its own blocks into the blockchain or split them off to create an independent branch that converges with the main blockchain at a later time.[13] The advantage for the attacker in carrying out such an attack is that he/she can double his/her own money and selectively reject transactions that he/she does not want to include in the blockchain.

Despite temporary forks within the Bitcoin blockchain, this system remains consistent within itself, but at the expense of transaction confirmation times. So the Bitcoin blockchain creates a maximum of 7 transactions per second—in comparison to, for example, Mastercard or Visa, which can handle more than 10,000 transactions per second.

It could also be criticized that, despite the use of the computing power of all participating nodes in the system, only one is always fast enough to solve the pending task and calculate the correct hash. In this case, there is no second winner—only the fastest receives the corresponding compensation for mining.

---

[10] Winner node in the sense of the node that has calculated the corresponding block.

[11] Fertig/Schütz (2019), pp. 88, 178.

[12] https://bitnodes.earn.com, accessed: 15.05.2022.

[13] Cointelegraph (2019): https://de.cointelegraph.com/news/two-miners-purportedly-execute-51-attack-on-bitcoin-cash-blockchain, accessed: 18.08.2019.

The computing power used by all participants demands its tribute in the form of a very, very high energy consumption. The Cambridge University therefore determines the "Cambridge Bitcoin Electricity Consumption Index" (CBECI), which is updated every 30 s.[14] At the end of July 2022, the electricity consumption is estimated at around 81 terawatt hours. The Technical University of Munich has also drawn up a calculation taking into account the equipment of the miners and comes to the conclusion that the mining of Bitcoin worldwide causes at least 22 million tons of CO_2—as much as, for example, the city of Hamburg with households and industry emits each year.[15]

Nevertheless, PoW protocols are important because they enable the construction of early blockchain networks. But they remain very inefficient.

2.2   Proof of Space (PoS)

Many discussions about consensus mechanisms and the way transactions should be conducted on a blockchain revolve around energy consumption. The Proof of Work mechanism (see Sect. 2.1) is repeatedly criticized for wasting energy. When in March 2021 the company Tesla announced that it would now accept Bitcoin as a means of payment,[16] it took only a few weeks for the founder to back down. At the end of June 2021, Musk does indeed express the possibility of accepting Bitcoin again, but only if the energy used in mining comes from at least 50% renewable energy.[17]

To counter this discussion, another—similar approach could be interesting: With Proof of Space, the focus is on storage and not on calculation as with Proof of Work.[18] A user reserves a certain amount of storage space on his/her computer. The program downloads a software that assigns a unique number to each section of this storage space. In the Proof of Space Network, these uniquely assigned numbers are selected at random. The computer to which the segment and thus the corresponding number belongs, validates the transaction.

Supporters of this approach point out that most users have unused storage space on their devices and that a much lower power consumption is required to validate the transactions.[19] Whether this approach will prevail is not yet recognizable in summer 2021,

---

[14] https://www.cbeci.org, accessed: 15.08.2019.

[15] Stoll/Klaasen/Gallersdörfer, (2019), accessed on 15.08.2019.

[16] e.g. Spiegel: https://www.spiegel.de/wirtschaft/unternehmen/bitcoins-autobauer-tesla-akzeptiert-digitalwaehrung-als-zahlungsmittel-a-6cac5116-8f24-4c44-8717-b0e87037f3f6; accessed July 04, 2021.

[17] Cointelegraph (2021): https://de.cointelegraph.com/news/elon-musks-lays-out-when-tesla-will-begin-accepting-bitcoin-payments-again, accessed July 04, 2021.

[18] In 2013, this approach was put forward by Dziembowski et al as an alternative to Proof of Work.

[19] Dziembowski et al. (2013) S. 2, accessed on 16.08.2019.

even if first examples and coins with this approach can be observed in the market (see, for example, Chia Coin, which in addition to Proof of Space uses Proof of Time).[20] At the same time, one can observe increased demand for hardware with a lot of storage space[21] (more than 10 Terabytes).

2.3   Proof of Stake (POS)

In order to counteract the high energy consumption and still ensure a consensus in the network, the Proof of Stake uses the shares held (stake) in the cryptocurrency (e.g. a so-called "native coin" like Peercoin, NXT, etc.) as the basis for validating the blocks.[22] No extra hardware is required to validate blocks. What is decisive is rather the amount of the "stake".[23] The more shares a stakeholder has in the corresponding and circulating crypto-currency, the higher the probability that this stakeholder can create a block.

Example

For example, if a participant in this system owns 5% of all coins in circulation, this participant has a 5% chance of validating a block. ◄

This approach also differs linguistically. So miners become *forgers* or *validators,* and blocks are *forged* instead of minted.[24]

The incentive to participate in this system is not through a rewards system like Proof of Work, where the miner gets the valid fee for each block minted. In Proof of Stake, the validators get the transaction fees incurred per block.

However, the PoS does not only bind the validators to the system through their share of currencies, but also through the reliability of the validators, as they "forge" blocks. Every validator has a very own interest in the security of the system in order to protect the value of his/her share. If a validator would validate unlawful transactions, he/she would lose his/her share (stake), which he/she deposited as a pledge, because he/she harms the system as a whole with the invalidation. However, the "allocation" of the block to be validated does not only depend on the amount of the held share, but rather a random algorithm distributes the blocks to be validated depending on the held share.

---

[20] Business White Paper by Chia Network Inc. (2021), S. 9 ff.

[21] Ingenieur.de (2021): https://www.ingenieur.de/technik/fachbereiche/ittk/kryptowaehrung-chia-werden-festplatten-jetzt-unbezahlbar-das-steckt-dahinter/, accessed on 04.07.2021.

[22] Morabito (2017), p. 11.

[23] Tapscott/Taspcott (2016), p. 32.

[24] Fertig/Schütz (2019), p. 145.

Thus, no validator can predict whether and when he/she will receive which block for validation.[25]

Background Information

Basically, two PoS approaches can be distinguished, on the one hand the "chain-based PoS system", on the other hand the "Byzantine fault tolerance (BFT) PoS system".

In the chain-based PoS, the algorithm randomly selects a validator during each time window (e.g. a time window can be determined within a period of e.g. 10 s) and assigns this validator the right to create a single block, and this block must point to a previous block (usually the block at the end of the previously longest chain), and so most blocks converge over time to a single, constantly growing chain.

In the BFT-PoS system, the validators are randomly assigned the right to propose blocks, but the agreement as to which block is valid is made by a multi-stage process in which each validator sends a "vote" for a certain block during each round, and at the end of the process all (honest and online) validators permanently vote on whether a certain block is part of the chain or not. Note that blocks can still be connected to each other; the main difference is that the consensus on a block can be within a block and does not depend on the length or size of the subsequent chain.

Even though the PoS approach has advantages over the PoW approach (e.g. less energy consumption than PoW), it is also not free of challenges. For example, it can be criticized that there is a minimum deposit in order to receive a reward. While this objection is easily refuted, because even with PoW a participant first has to invest in hardware, but a too small stake will make it almost impossible to ever validate a block. This then leads to possible unfair conditions, since the lottery character inherent in the system only comes into effect with sufficiently large stakes. The risk of a monopoly, or at least an oligopoly, cannot be denied, since participants with a high stake receive the opportunity to validate a block much more often. These participants earn more from transaction fees than others with a low stake. In order to avoid this, the mere share in the stake is no longer sufficient for the traditional PoS systems to validate blocks. The random selection (randomized block selection) is already described above. The age of the coins can also be used to bring more fairness into the system. The older the coin is, i.e. the longer a validator holds it, the higher the probability of validating a block. For this purpose, a node must hold its coins for at least 30 days in order to even appear as a potential forger. If this node has validated a block, the waiting time begins anew.

In Delegated Proof of Stake (dPoS), so-called witnesses and delegates are elected who take on certain tasks in the network. This can be the validation of a block, but also adjustments to the consensus, for which a reward can be paid. The witnesses and delegates are always deselectable. An extension of this procedure can be found in the randomized Proof of Stake, in which a committee is determined to fulfill the tasks. The

---

[25] Bogensperger et al. (2018), p. 145.

random variable selects the members of the committee.[26,27] With these different additions to Proof of Stake, the obvious disadvantages should be eliminated to ensure more participation.

Vitalik Buterin, the founder of Ethereum, already published his/her understanding of these two systems in an essay in 2016. He explains that the capital invested guarantees safe and correct behavior, while the hardware and energy expenditure used for the credibility of the validators is used in PoW.[28]

Background information

Vitalik Buterin has been expressing himself for some time about taking the blockchain Ethereum, which he founded, to a new level with far higher transaction rates than before, while at the same time switching from PoW to PoS. His focus is on keeping the Ethereum blockchain decentralized while at the same time ensuring a low entry barrier for network validators. The success of Ethereum is also a burden, because the more applications are offered and validated over this blockchain, the slower the validation will be, because the network only has a certain capacity. For this reason, it is important to drive development forward in order to validate more transactions safely and quickly, but with much less energy consumption.

The programming code that drives the switch and bears the name "Phase Zero" has been available since 30.06.2019. It is assumed that it will take until at least the mid of 2022 before the switch from PoW to PoS is made at Ethereum. In December 2020, Ethereum 2.0 went online for the first time. This is the so-called Beacon Chain, which serves as the basis for the next necessary steps in the integration of the existing Ethereum network.[29]

2.4   Proof of Believability (POB)

This proof is still a relatively young and therefore not yet widely used approach in comparison to the others. The mechanism working in the background is based on the examination of the previous behaviour and the contributions of each node in the network. IOST[30] developed this approach in 2018 and promises better performance than Proof-of-Stake. Even though each individual application case or each company has to check which consensus mechanism fits the question at hand during the corresponding process optimizations, it is definitely worth taking a closer look at this proof for your own project.

Blocks and transactions are validated with this Proof-of-Believability by a "committee" deciding on the validity. The committee currently consists of 17 nodes and every

---

[26] Fertig/Schütz (2019), S. 147.

[27] Casey/Vigna (2018), S. 90.

[28] Buterin (2016) A proof of Stake Design Philosophy.

[29] https://ethereum.org/en/eth2/beacon-chain/, accessed on 04.07.2021.

[30] https://iost.io, accessed on 22.07.2021.

member of the community can participate if they have the community's own tokens called Servi (comparable to Ether at Ethereum). The composition of the committee is changed every ten minutes by selecting the 17 nodes with the highest Servi account balance. As long as the round is running, the selected committee now produces blocks and receives the corresponding reward. The peculiarity of this mechanism is that at all nodes representing the new committee, the Servi balance is reduced by the balance of the 17th node. This means: The Servi of the 17th node "burns" its tokens or the balance is reset to zero and at the other 16 nodes the balance is reduced by exactly the amount that the 17th node has given.

Example: the 17 selected nodes all have more than one Servi. Node No. 17 has two Servi. The account balance in Servi at the beginning of the validation round is thus zero and all other 16 nodes also have two Servi deducted from their account balance. This means that the nodes not selected in the current round have partially more Servi, which means that they can be selected for the committee in the next round. This procedure makes it possible for the composition of the committee to change constantly. The probability of validating transactions and receiving the reward for this using this mechanism is much higher than with Proof-of-Stake.[31]

2.5   Delegated Byzantine Fault Tolerance (dBFT)

The "delegated Byzantine fault tolerance" (dBFT) approach is a demanding algorithm that is intended to facilitate consensus over a blockchain.

The background to this solution approach is the question from game theory that has found its way into computer science under the name "problem of the Byzantine generals".[32] This dilemma from game theory shows the communication problems a group of generals is confronted with in the event of a strategically important joint attack.[33]

The dilemma assumes that each general has his/her own army, which was camped in different places around, for example, a city. The generals have to agree on an attack or retreat. In addition, a decision once made can not be undone, and the decisions of the individual generals must be carried out at the same time. The difficulty is that one or more of the generals could be a traitor, which in turn means that they could make false statements about their procedure. To make matters worse, the communication between the generals takes place only by couriers. An attack then fails if there is no consensus of action.[34]

---

[31] https://iost.io/1193/ accessed on 22.07.2021.

[32] Bahga/Madisetti (2017), p. 352.

[33] Holler/Illig/Napel (2019), pp. 24 ff.

[34] Akkoyunly, E.A./Ekanadham K./Huber R.V (1975), p. 70 ff.

Transferred to the blockchain world, this means that the "generals" are the nodes in the system. The majority of the nodes within the distributed network must agree on and carry out the same action in order to maintain a secure and stable network. At least 2/3 approval of reliable and honest nodes is required. The (Byzantine) fault tolerance in this context means that the system continues to work even if up to 1/3 of the nodes fail or act maliciously.[35]

As with delegated Proof of Stake, this approach has in common that it assumes that the participants of such a network have an intrinsic interest in interacting in an honest system. If delegates are now determined (delegated Byzantine Fault Tolerance), then those are elected who are considered honest enough by the users to maintain the integrity of the blockchain network.

Fertig and Schütz describing the roles that delegates can take on in this system as follows:[36]

- *Consensus Node:* These delegates are elected by the entire network, and their task is to ensure compliance with the consensus.
- *Speaker Nodes* create the blocks.
- *Delegate Nodes* validate the blocks and the transactions contained therein.

The Chinese company Neo, a blockchain platform for developing digital assets and integrating smart contracts, uses this approach. Interesting is the statement of the company to be able to carry out more than 10,000 transactions per second in the future. This brings the blockchain closer to the transaction speed of traditional databases.[37]

Although not yet as widespread, this approach provides an alternative with a simpler demonstration of the use of work methods.

2.6   Proof of Authority (PoA)

Stronger than the dBFT approach, Proof of Authority is particularly widespread in Private Permissioned Blockchain. This approach is a variation of Proof of Stake. The validators are usually named by the operators of the system as authorities before commissioning the blockchain. These authorities do not have to hold a stake to create and validate blocks. The credibility of selected nodes as authorities ensures the smooth operation within this blockchain. So authorities can exclude nodes that violate the system's rules from the system.

---

[35] Lamport/Shostak/Peace (1982), p. 385 ff.

[36] Fertig/Schütz (2019), p. 147.

[37] Neo Whitepaper, Consensus Mechanisms, accessed 10/09/2019.

According to the Proof-of-Authority algorithm, it is determined in turn among the authorities who is the "mining leader" and thus has the right to propose new blocks. The majority of the other authorities must confirm the new block in order for it to be attached to the previous blocks.[38]

Since Proof of Authority does not require mining or any other (native) currency for processing, this approach consumes little energy and computing power to validate the blocks. Transactions can be processed faster and more efficiently than, for example, with Proof of Work. This increases scalability. The authorities ensure compliance with the consensus.[39]

Because the authorities play such a prominent role as validators, these must be protected as nodes against attacks and other manipulation attempts in order to maintain security.

Overall, however, this is a more centrally oriented approach, and thus this approach is well suited for corporate solutions, but less or not at all for public blockchain approaches.

2.7   Proof of Activity (PoAc)[40]

In this approach, the two best-known mechanisms (PoW and PoS) are linked together. So the Proof of Activity starts with the mining process as the standard POW process, in which different miners try to outbid each other with higher computing power to find a new block. If the new block is found, which only contains a header and the reward address of the miner, the system switches to the PoS approach. Based on the header, a new random group of auditors is selected from the blockchain network, which must validate or sign the new block. The more stake a validator has, the more chances he/she or she has to be selected as a signatory.

Once all validators have signed the newly found block, it receives the status of a complete block, is identified and added to the blockchain network, and transactions are recorded.

An attack on this system appears to be almost impossible, because in addition to the so-called 51% attack, which states that an instance must have more than 50% of the world's hashing power, an attacker would also have to have the majority of the coins of the respective blockchain as stake.

However, this advantage turns into a disadvantage, because mining causes high energy costs and PoS promotes monopoly formation.

---

[38] De Angelis et al. (2017), p. 2.
[39] Fertig/Schütz (2019), p. 150.
[40] Bentov et al. (2014).

2.8   Proof of Importance (PoI)

This proof mechanism is also derived from proof of stake and is known through the NEM blockchain platform. Proof of importance is therefore the mechanism by which it is determined which network participants (nodes) are entitled to add a block to the blockchain. A prerequisite for this is that potential validators have a sufficient number[41] of cryptocurrency in custody to qualify. In addition to the stake, productive activity in the network is included.[42]

The proof of importance can be seen as a new consensus algorithm, because it, unlike existing consensus mechanisms such as PoS, tries to take into account the general support of the network.

2.9   Proof of Reputation (PoR)

The reputation of a participant, is of fundamental importance for this consensus mechanism. Reputation can therefore be defined as the evaluation of the trustworthiness of a member by others. In peer-to-peer networks, reputation systems can be used because they enable the ability to trust each other and thus allow successful interactions.[43] Those participants who have a good reputation can be given the right to validate and create blocks.

The GoChain project uses this approach and generally argues that neither miners nor coins are needed to ensure the security of the underlying blockchain. Rather, a participant with sufficient reputation would be confronted with significant financial and trademark consequences if he/she or she attempted to cheat the system. This is a relative concept, as almost all companies would suffer greatly if they were caught trying to act fraudulently. However, larger companies usually have more to lose.[44]

This can also make PoR attractive for public blockchains, because large companies lose too much in the event of unethical behavior. In addition, potential new participants can take a look at the structure of the network and decide accordingly whether they trust the validators.[45]

---

[41] According to the NEM protocol, an account must have at least 10,000 unredeemable XEM (i.e. the currency of the NEM platform) to be eligible for this.

[42] Fertig/Schütz (2019), p. 149.

[43] Gai et al. (2018), p. 667.

[44] GoChain   (2019)   Medium   https://medium.com/gochain/proof-of-reputation-e37432420712; accessed 15.August 2019.

[45] Fertig/Schütz (2019), p. 151.

2.10   Proof of Elapsed Time (PoET)

Chip manufacturer Intel developed this proof to reduce energy consumption and other resource usage compared to, for example, PoW. The consensus mechanism makes this process particularly well suited for permissioned blockchain approaches. In networks using this approach, the participant must identify themselves in advance before they can join.

The PoET algorithm works as follows: Each participating node in the network must wait for a randomly chosen period of time, and the first one to fulfill the specified waiting time wins the new block. Each node in the blockchain network generates a random waiting time and goes into standby mode for the specified time. The one who "wakes up" first—that is, the one with the shortest waiting time—transfers a new block to the blockchain by sending the necessary information to the entire peer network. The same process is then repeated for the discovery of the next block.

Based on the principle of a fair lottery system, in which each individual node has an equal chance of being a winner, the PoET mechanism is based on distributing the chances of winning to as many network participants as possible. The first participant who ends the wait will be the leader for the new block.

For this to work, two requirements must be checked. First: Did the lottery winner actually choose a random waiting time? Otherwise, a participant could deliberately choose a short waiting time to win. Second: Did the lottery winner actually wait the specified time?

The built-in mechanism allows applications to run trusted code in a protected environment, while ensuring that both requirements—for the random selection of the waiting time for all participating nodes and the actual fulfillment of the waiting time by the successful participant—are met.[46]

2.11   Proof of Burn (PoB)[47]

This approach also aims to waste less resources than, for example, PoW, although it partially simulates this.

In Proof of Burn, coins are "burned" to permanently remove them from the network. For this purpose, an address (an so-called Eater Address) is predetermined to which the coins to be destroyed are to be sent. This burn address is known so that all network participants can verify the destroyed and removed coins. The participant who has sent

---

[46] Hyperledger Sawtooth: https://sawtooth.hyperledger.org/docs/core/releases/1.0/architecture/poet. html, accessed: 15. August 2019.

[47] Fertig/Schütz (2019), S. 152.

his/her coins to this address to have them burned receives in return the right to validate blocks, that is, his/her source of income in the event of validation.

As already described in the PoS approach, the validation of blocks in Proof of Burn is subject to the share (in this case) of the destroyed coins of the individual participant as well as the probability associated with it: The more coins a participant destroys, the higher his/her probability of receiving new blocks for validation. The built-in time delay also means that a participant who destroys a share of his/her coins must wait a certain period of time before he/she is allowed to validate. This ensures that a participant cannot validate the block in which he/she has burned his/her own coins. After the expiration of the time period specified in the protocol, the participant can participate in the "lottery" of the blocks.

The idea behind Proof of Burn is to retain the concept of investing resources in the blockchain while reducing the need to invest in external resources such as intensive electricity consumption. The "burning" of coins solves this problem and also creates a degree of built-in scarcity for the coins. Furthermore, the power of the burned coins "decays" or decreases whenever a new block is mined. This ensures the regular activity of miners in the network instead of allowing them to have constant chances of validating blocks with a one-time investment. In PoB, this mechanism can be used to establish balance in the network: On the one hand, miners in Proof-of-Burn networks are still compensated in coins for their efforts in order to reward both the miner and to prevent the number of coins from falling below a certain level. On the other hand, transaction fees are charged, which the miner receives who has validated the block.

This approach is seen as advantageous in that a Proof-of-Burn protocol promotes long-term involvement in a project or corresponding network. If there is a higher percentage of long-term investors, the price of coins could be more stable. Also, the proof of burning helps to determine the distribution of the cryptocurrency fairly and decentralized.

It is criticized, however, that the proof of combustion wastes resources, similar to proof of work. In addition, those miners who are willing to burn more money receive power. This is a problem similar to proof of stake, where it is argued that the rich get richer because they have a higher chance of validation with more shares.

2.12   Zero-Knowledge-Proof (ZKP)[48]

Transparency is perceived as one of the important properties of blockchain solutions. For one or the other application, however, this transparency could be rather disadvantageous. There are companies, for example in the financial sector, which deal with sensitive

---

[48] Altoros   (2019)   https://www.altoros.com/blog/zero-knowledge-proof-improving-privacy-for-a-block-chain/, accessed: 17. August 2019.

information. Privacy may take precedence over transparency. For companies that work with confidential information, the implementation of blockchain transactions with zero-knowledge proof may be the solution.

ZKP is a method in cryptography in which a so-called prover (a participant within the network) convinces another participant, called verifier, that he/she knows a secret value without disclosing the actually relevant information. The prover explains that he/she knows the secret.

The essence of a ZKP is to prove in a simple way that someone has knowledge of certain information without revealing it. ZKP are not proofs in the mathematical sense of the word, because there is a small probability that a fraudulent prover will be able to convince the verifier of a false statement. ZKP are therefore more probabilistic than deterministic proofs.

Three essential features for a Zero-Knowledge-Proof can be identified:

- *Completeness:* If the statement a prover makes is true, it will also convince the honest verifier.
- *Soundness:* If the statement of the prover should be false, it will not convince the verifier.
- *Zero-Knowledge:* If the statement is true, the verifier will not receive any information about the content of the statement, but only that the statement of the prover is correct.

The general course of a Zero-Knowledge-Proof consists of three consecutive actions between the participants A (prover) and B (verifier). These actions are referred to as testimony, challenge and answer.

- *Testimony:* The fact that A knows the secret determines a part of the questions that can always be answered correctly by A. First, A randomly chooses any question from the set and calculates a proof. Then A sends the proof to B.
- *Challenge:* Afterwards, B chooses a question from the set and asks A to answer it.
- *Answer:* A calculates the answer and sends it back to B.

The received answer allows B to check whether A really knows the secret.

This process can be repeated any number of times to make the probability that A is merely guessing, rather than knowing the right answers, small enough to satisfy B.

Applications that benefit from ZKP are those that require a certain degree of privacy, such as:

Authentication systems: The development of ZKP was inspired by authentication systems in which one party had to prove its identity to a second party using some secret information without revealing the secret completely.

Anonymous systems: ZKP can enable the validation of blockchain transactions without revealing the identity of the users who perform the transaction.

Confidential systems: Similar to anonymous systems, ZKP can be used instead to validate blockchain transactions without revealing relevant information such as financial details.

2.13 Ripple

Ripple is a digital currency and a payment system that was developed entirely without any dependence on Bitcoin. It is independent of any mining protocol and has a publicly disclosed ledger.

Ripple's technology has accomplished some new things. There are no miners. Instead, a consensus algorithm is used that relies on trusted subnetworks to keep a broader decentralized network of validators in sync. It is important to note that the Ripple consensus algorithm relies on trust and is significantly different from Bitcoin's proof-of-work design.

In the Ripple system, each validating server keeps a list of trusted servers called a Unique Node List (UNL), and each server only trusts the votes cast by the servers that are listed in the UNL.

Each validating server in the Ripple system authenticates any proposed changes to the last closed ledger, which is also referred to as the most recently validated ledger, and the changes that are agreed to by half or more of the servers are incorporated into a new proposal that is then sent to the other servers in the network. This process is then repeated, with the voting requirements being raised to 60% of the servers, then 70% of the servers, and then 80% of the servers. After this process, the server then authenticates the changes before informing the network of the completion of the most recently validated ledger. Any transaction that originally took place but was not included in the ledger is discarded at this time. Such discarded transactions are then considered invalid by users of the Ripple system.[49]

Ripple uses these trusted gateways as endpoints for users, and these gateways could accept deposits and redeem debts in all forms of assets, including traditional fiat currency.

2.14 Summary

Table 2.1 picks up the discussed proof concepts including their properties and thus allows for a comparison. This knowledge about these concepts is helpful when it comes to examining whether a company needs a blockchain solution (or not). The properties mentioned here play a role, among other things, in the incentive for potential participants

---

[49]Morabito (2017), p. 96.

Table 2.1 Consensus mechanisms

	PoW (Proof of Work)	PoS (Proof of Stake)	delegated Proof of Stake	PoET (Proof of Elapsed Time)	BFT (Byzantine Fault Tolerance) and variants	Ripple
Blockchain Type	Public	Private (restricted access) & public		Private (restricted access) and public	Private (restricted access)	Private and public
Transaction completion	Probability theory	Probability theory		Probability theory		
Transaction speed	Low, approx. 3.5–7 tx/sec	Low, approx. 15 tx/sec		Medium	High	Medium
Energy expenditure	High	Medium		Low	Medium	Low
Scalability of the peer network	High	High		High	Medium	High
Opposing Tolerance	<51% of processing power	<51% of capital employed and dependent on algorithm		Unknown	<33% of replicas are faulty	<20% of faulty nodes in UNL
Objective	To create a barrier to entry for publishing blocks in the form of a computationally difficult puzzle to allow transactions between untrusted participants	To create a less computationally intensive (compared to PoW) barrier to entry for publishing blocks while still allowing transactions between untrusted participants	To enable a more efficient consensus model through a "liquid democracy" in which participants (using cryptographically signed messages) vote to elect and revoke the rights of delegates to validate and secure the blockchain	To allow for a more economical consensus model for blockchain networks, but at the expense of deeper security guarantees associated with PoW, as the consensus is centralized	To ensure a functioning blockchain network even if some of the nodes fail or act maliciously	To quickly and inexpensively ensure cross-border payment transactions

(continued)

Table 2.1   (continued)

	PoW (Proof of Work)	PoS (Proof of Stake)	delegated Proof of Stake	PoET (Proof of Elapsed Time)	BFT (Byzantine Fault Tolerance) and variants	Ripple
Advantages	Attacking the network (denial of service) is difficult to implement because the blockchain network is decentralized. The decentralized nature of the blockchain means that it can theoretically allocate data and bandwidth to absorb DDoS attacks if they occur. Open to anyone who has the right hardware to solve the mathematical puzzle.	Less computationally intensive than PoW. Open to anyone who wants to use cryptocurrencies. Stakeholders control the system.	Elected delegates are economically motivated to stay honest. More computationally efficient than PoW.	Far less computationally intensive than PoW.	Securing the blockchain network through BFT.	It is a bank-independent currency that seems to be universally applicable.

(continued)

Table 2.1 (continued)

	PoW (Proof of Work)	PoS (Proof of Stake)	delegated Proof of Stake	PoET (Proof of Elapsed Time)	BFT (Byzantine Fault Tolerance) and variants	Ripple
Disadvantages	Intensive computing (by design), power consumption, hardware arms race. Potential for 51% attack through sufficient computing power.	Stakeholders control the system. Nothing can prevent the formation of a pool of interest groups to create a centralized power. Potential for 51% attacks through sufficient financial power exists.	Less node diversity than PoW or pure PoS consensus implementations. Greater security risk for node compromises by limited set of operational nodes. Since all delegates are "known", there may be an incentive for block producers to join forces and accept bribes, which would jeopardize the security of the system.	Given the late latency limit, true time synchronization in distributed systems is essentially impossible. PoET requires Intel Software Guard Extensions (SGX).	Limited scalability.	All Validator Nodes are operated by Ripple itself. Thus, high influence by the company Ripple on the network; centralized approach.
Example	Bitcoin	Ethereum, NAVCoin, Neo, Lisk etc.	Bitshare; STEEM, Cardano	Hyperledger Sawtooth	Hyper Ledger Fabric, Neo	Ripple

(i.e. how to motivate future participants in the course of one's own solution to get involved). The question of a private or perhaps public blockchain can also be answered better in the context of the properties.

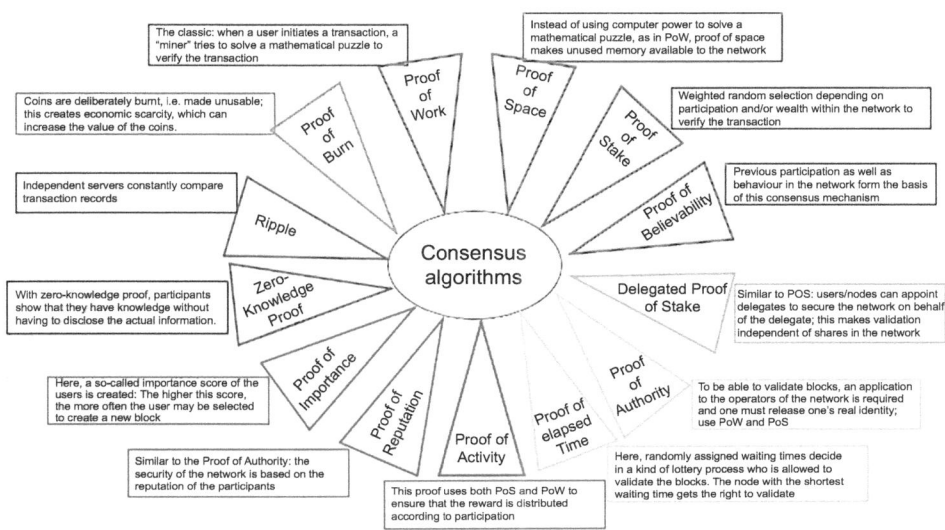

Fig. 2.2 Consensus models

It should be noted at this point that there are more than the approaches listed here. This selection is based on the personal assessment of the author with regard to their importance. This knowledge is used in Sect. 4.9., when it comes to working out a so-called framework for interesting processes. The summary presented here can help to accomplish the design of the framework. But the list of proof concepts is not exhaustive. Here a new field of research is emerging, which aims to determine the optimal consensus logic (cf. Table 2.1)

The mechanisms can be summarized graphically as shown in Fig. 2.2.

2.15   Conclusion

The large number of different consensus models gives the first impression of the wealth of design options. Proof of Work works by having all participating nodes solve cryptographic puzzles. This requires a lot of computer power to validate blocks. In addition to a certain amount of time, this now costs a lot of energy. Without the appropriate and more expensive hardware, it is almost impossible to validate a block. The advantage of this approach is the protection against hacker attacks: A successfully carried out attack must take place simultaneously at least at more than 50% of all participating nodes. With a large network like that of Bitcoin with more than 9400 nodes, such an effort is not possible with the currently conventional computers.

Despite strengths and weaknesses, this first proof approach has led to the development and introduction of new concepts. Each of these further approaches has its own advantages and disadvantages. Therefore, you not only have to decide which type of blockchain (public, private or consortium) you want to create, but also which rules you want to work with on this blockchain.

References

Akkoyunly EA, Ekanadham K, Huber RV (1975) Some contraints and tradeoffs in the design of network communikation. State University of New York, New York

Altoros   (2019).   https://www.altoros.com/blog/zero-knowledge-proof-improving-privacy-for-a-block-chain/. Accessed: 17. Aug 2019

de Angelis S et al (2017) PBFT vs proof of authority: applying the cap theorem to permissioned blockchain Research Center of Cyber Intelligence and Information Security, Sapienza University of Rome & University of Southampton

Bahga A, Madisetti V (2017) Blockchain application, a hands-on-approach. Book-On-Demand, Norderstedt

Bentov I et al (2014) Proof of activity: extending Bitcoin's proof of work via proof of stake. https://eprint.iacr.org/2014/452.pdf. Accessed: 10. Sept 2019

Bitnodes (2019). https://bitnodes.earn.com. Accessed: 16. Nov 2019

Blockchain Demo (2019) Anders.com. https://anders.com/blockchain/. Accessed: 5. Aug. 2019

Bogensperger A, Zeiselmair A, Hinterstocker M (2018) Die Blockchain Technologie: Chance zur Transformation der Energie-Versorgung? Forschungsstelle für Energie-Wirtschaft e. V. (FFE), München

Buterin V (2016) A proof of stake design philosophy, medium. https://medium.com/@VitalikButerin/a-proof-of-stake-design-philosophy-506585978d51. Accessed: 8. Aug. 2019

Cambridge University mit dem Cambridge Bitcoin Electricity Consumption Index. https://www.cbeci.org. Zugegriffen am 15.08.2019

Casey MJ, Vigna P (2018) The truth machine, the blockchain and the future of everything. HaperCollins Publishers, London

Chia Network Inc. (2021) Business white paper, version 2 Q.1. https://www.chia.net/2021/02/10/chia-businesss-whitepaper.html. Accessed: 20. July 21 und 24. Sept 2021

Cointelegraph   (2019).   https://de.cointelegraph.com/news/two-miners-purportedly-execute-51-attack-on-bitcoin-cash-blockchain. Accessed: 18. Aug 2019

Cointelegraph   (2021).   https://de.cointelegraph.com/news/elon-musks-lays-out-when-tesla-will-begin-accepting-bitcoin-payments-again. Accessed: 4. July 2021

Dziembowski S, Faust S, Kolmogorov V, Pietrzak K (2013) Proof of space, verfügbar unter ia.cr/2013/796 bzw. https://eprint.iacr.org/2013/796

Fertig T, Schütz A (2019) Blockchain für Entwickler, Grundlagen, Programmierung, Anwendungy. Rheinwerk, Bonn

Froystad P, Holm J (2015) Blockchain: powering the internet of value (White Paper). Evry Labs

Gai F et al (2018) Proof of reputation: a reputation-based consensus protocol for peer-to-peer-network, in databased systems for advanced application. Springer International, Cham

GoChain (2019) auf Medium. https://medium.com/gochain/proof-of-reputation-e37432420712. Accessed: 15. Aug 2019

Hashgenerator (2019). https://passwordsgenerator.net/sha256-hash-generator/. Accessed: 18. Nov 2019

Holler MJ, Illing G, Napel S (2019) Einführung in die Spieltheorie, 8. Aufl. Springer Gabler, Berlin

https://ethereum.org/en/eth2/beacon-chain/. Accessed: 4. July 2021

https://iost.io. Accessed: 22. July 2021

Hyperleder Sawtooth. https://sawtooth.hyperledger.org/docs/core/releases/1.0/architecture/poet.html. Accessed: 15. Aug 2019

Ingenieur.de (2021). https://www.ingenieur.de/technik/fachbereiche/ittk/kryptowaehrung-chia-werden-festplatten-jetzt-unbezahlbar-das-steckt-dahinter/. Accessed: 4. July .2021

Lamport L, Shostak R, Pease M (1982) The Byzantine generals problem. ACM Trans Program Lang Sys 4(3):382–401

Morabito V (2017) Business innovation through blockchain, The B^3 Perspektive. Springer International, Cham

Neo Whitepaper. https://docs.neo.org/docs/en-us/basic/whitepaper.html.Accessed: 10. Sept 2019

Spiegel. https://www.spiegel.de/wirtschaft/unternehmen/bitcoins-autobauer-tesla-akzeptiert-digital-waehrung-als-zahlungsmittel-a-6cac5116-8f24-4c44-8717-b0e87037f3f6. Accessed: 4. July 2021

Stoll C, Klaaßen L, Gallerdörfer U (2019) The carbon footprint of Bizcoin. https://doi.org/10.1016/j.joule.2019.05.012

Tapscott D, Tapscott A (2016) Blockchain revolution, how the technology behind bitcoin is changing money, business, and the world. Penguin Random House, New York

Further Elements in the Blockchain System

3

Abstract

This chapter describes the other important components of blockchain technology. In addition to the cryptographic basics, this also includes an understanding of smart contracts, which, programmed as self-executing contracts, enable the execution of credible transactions without third parties.

You have learned the basics and can distinguish between the different approaches and consensus mechanisms. The blockchain system includes other important components that need to be considered in more detail in order to be able to assign the context. This chapter describes cryptography and smart contracts and gives an overview of the context in which this technology is tested or applied. Of course, reference is made to the first large and successful application case—digital currency. In addition, the system-immanent security aspects are briefly illuminated. Blockchain solutions must create value—but this value is not equally easy to define for everyone. There will always be an individual assessment.

3.1 Cryptography

Another important part of the blockchain is cryptography, a field of computer security that is widely used. Satoshi Nakamoto also made use of this and transferred it to his system Bitcoin/Blockchain. The term "crypto" comes from the Greek and means "secret" or "hidden", and "-graphy" means "writing". Together it is the "secret writing" or the "encrypted writing". If a text is encrypted, there must be a key to make the text readable.

© The Author(s), under exclusive license to Springer-Verlag GmbH, DE, part of Springer Nature 2022
K. Adam, *Blockchain Technology for Business Processes*,
https://doi.org/10.1007/978-3-662-65818-5_3

Table 3.1 Caesar Code or ROT13

A	B	C	D	E	F	G	H	I	J	K	L	M	N	O	P	Q	R	S	T	U	V	W	X	Y	Z
N	O	P	Q	R	S	T	U	V	W	X	Y	Z	A	B	C	D	E	F	G	H	I	J	K	L	M

The desire to keep information secret from unauthorized persons is at least as old as writing itself. The first areas of application are in the military, whether in ancient Egypt or Rome. A famous example of a simple, symmetrical encryption code is the so-called *Caesar code*, in which each letter in the alphabet is replaced by a linear displacement. ROT13 (Caesar encryption by 13 places) is a widely used encryption code that, extended to Tripple-ROT13, also meets more demanding security criteria.

Table 3.1 shows the simple ROT13 code with a displacement of 13 places.

ROT13 code is a halving of the alphabet, and thus encryption and decryption are the same process.[1]

In connection with blockchain technology, a number of terms are mentioned here, which are defined briefly:

In addition to symmetrical encryption, as with the Caesar code, where the same key is used for encryption and decryption, one also speaks of asymmetrical encryption. For this purpose, a so-called *public-private key pair* is created. For encryption, a different key is therefore used than for decryption. The *public key* can be distributed to anyone; it is public. This means that, for example, if person A wants to write a message, they can use A's public key. A will decrypt the sent message with his/her private key. This is comparable to your e-mail inbox. Your e-mail address is known, anyone who wants to can write you a message. But as the recipient, you can only decrypt this message with your private key. For this purpose, e-mail encryption uses solutions such as *Pretty Good Privacy* (PGP) with asymmetric key pairs.[2]

The term digital signature is also often used in connection with blockchain. The range extends from electronic signatures in the sense of an electronic signature (simple) to digital signatures. A digital signature is certified by a trusted certification authority and has a very high security standard. These signatures are electronic fingerprints and enable signing as well as authenticating the signatory.

In the context of blockchain, it does not seem (necessarily) necessary to use a digital signature as a certified message (or transaction), because this is actually implemented in the blockchain system. A blockchain maps by means of the consensus mechanism when and under which conditions a transaction is valid. No certification by a central (trusted) instance is required for this. The network decides. Nevertheless, the blockchain uses asymmetric encryption methods with a so-called one-way function to ensure

---

[1] Fertig/Schütz (2019), p. 68.
[2] Kersken (2019), p. 1287.

authentication.[3] For this, a user chooses a randomly created private key from, for example, 2^{256} possible number combinations.[4] This allows the calculation of a public key. This is a so-called one-way function, because it is almost impossible to calculate the private key back from the value of the public key.[5]

If A now sends a message, he/she encrypts it with his/her private key. The message can only be decrypted with a public key that is calculated from A's private key. This allows it to be reliably demonstrated that the message (or transaction) actually comes from A (and no one else). A "nice side effect" is that the falsification of the message (transaction) is also prevented. For a counterfeiter it is almost impossible to encrypt the message again so that the "correct" (public) key can be derived.

Background Information
The creation of key pairs usually takes place on a blockchain according to the so-called "Elliptic Curve Cryptography" (ECC) and uses the special features of elliptical curves. This procedure was independently presented in 1985 by Neal Koblitz and Victor Miller. It is a procedure in which asymmetric cryptographic methods are calculated on elliptical curves over finite fields. It is important to note that elliptical curves are not ellipses. They are calculated to the square. Elliptical curves are calculated as a cubic number (x^3).
Mathematically expressed, the formula looks as follows:[6]

$$y^2 = x^3 + ax + b(\mathrm{mod\,p})$$

Graphically, the solution set is shown in Fig. 3.1 under assumptions
p = -3 and p = 3 as shown in Fig. 3.1.
Properties of these curves include, but are not limited to, that they always run horizontally symmetrical to the x-axis and that a non-vertical line intersects the curve at a maximum of three points. Thus, on curves of this type, a line can be drawn by adding points that can meet the curve at infinity.
These properties are used to generate asymmetric key pairs in the form of public and private key. A main advantage of using elliptic, curve-based cryptography is the reduced key size and thus increased speed.
Antonopoulus explains that for cryptographers, the elliptic curve discussion is a kind of "trap door":
The owner of a private key can easily create the public key and then share it with the world, knowing that no one can reverse the function and calculate the private key from the public key. [7]
The "Standard for Efficient Cryptography" institute is continuously working on the further development of these approaches, and deeper literature can be found under http://www.secg.org.[8]

---

[3] Fertig/Schütz (2019), p. 70.

[4] Antonopoulos (2017), p. 58.

[5] One-way functions are functions that are easy and easy to calculate in one direction (public key). The "back-calculation" as an inverse function is, however, very difficult and only possible with great effort.

[6] Antonopoulos (2017), p. 61.

[7] Antonopoulos (2017), p. 60.

[8] Standard for Efficient Cryptography: http://www.secg.org; accessed October 16, 2019.

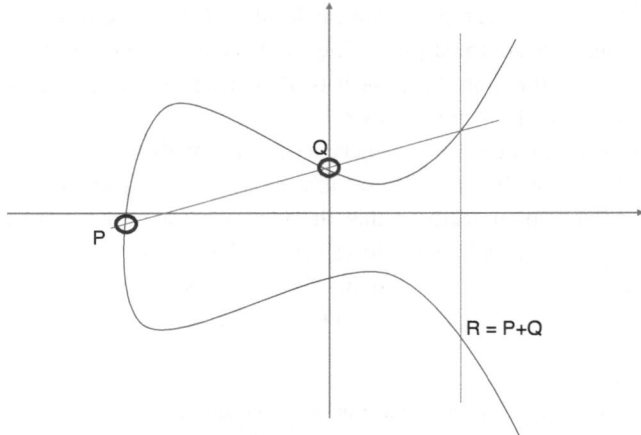

Fig. 3.1 Solution set of an elliptical curve

3.2   Smart Contracts

As already explained in Sect. 1.1, blockchain technology is also not stopping its development. With blockchain 1.0 and the application case of Bitcoin, it is possible to settle transactions of digital currencies. This is a so-called first-to-file application, where the order is of crucial importance. However, these are "only" payment transfers from one digital wallet to another. Values or, in new German, assets could not be transferred in the first version. In December 2013, Vitalik Buterin published his idea of an extension for the use of blockchain technology (and technique). In his "Ethereum White Paper: A Next Generation Smart Contract & Decentralized Application Platform", he picks up the approach of the Bitcoin Blockchain and explains how, in addition to the transfer of currency units, other assets such as land registers, medical records, trade register entries, identity cards, etc. can be stored reliably and securely within a blockchain network.[9] This means that completely new, automated business models based on peer-to-peer exchange are possible.

This approach is also not new. As early as 1997, Nick Szabo developed and described the concept of "secure property rights with owner authority". The basic idea is a new, replicating database technology that, similar to the blockchain system, makes it possible to store registers.[10] However, at the time of this concept, there is still no secure, replicating database, so this approach has not yet been implemented in practice.

---

[9] Buterin (c) (2013), p. 14.
[10] Szabo (1997).

Despite this, Nick Szabo defines the term as follows:

"Smart Contracts are computer protocols that map contracts or verify them or support nego-
tiation or execution."

From a technical point of view, a smart contract comes into effect as a "if-then" process. This can be triggered by an external event or by a user, provided that the pre-
defined conditions are met.[11] In this way, contracts between the contracting parties can
be mapped as a computer protocol by means of a corresponding programming language.
The conditions specified in the contract are automatically carried out, checked and fulfilled. All nodes involved in the blockchain network have this contract on their computers
and can thus also trace the verifiability and enforceability.[12]

The execution of such an automated smart contract can take place in three variations.
The easiest way to explain this is with the example of a vending machine. The mechanism of such a programmed machine is very similar, albeit simplified, to the smart-con-
tact approach from the blockchain.

Let's say you want to buy a drink from this machine. You put your coin in the corresponding slot and press a corresponding key combination to get the drink of your choice.
This is an so-called external event that internally sets a process in motion in the machine.

Variant 1: You have paid the required amount of money, pressed the corresponding
key combination and, since the machine is filled correctly, you receive your desired
drink. So everything went as desired and you as a buyer are satisfied. So the smart contract is fulfilled according to the specifications.

Variant 2 is the same in terms of setting: you pay the required amount, press the corresponding key combination and—receive nothing because the shaft you have selected
on the machine is empty. You are annoyed that you did not get a drink, press the refund
button and get your money back. Now the smart contract has not been fulfilled, but you
have got your money back, so no damage has been done (apart from the fact that you
could not quench your thirst).

In variant 3 it gets interesting. Let's assume again that you pay, press the desired key
combination and receive nothing. The shaft may be empty or a bottle may be jammed
in the machine. Regardless of the cause: you do not get your desired product and the
contract is not fulfilled. In addition, you do not receive—for whatever reason—your
money back, even though you pressed the corresponding button. From a programming
point of view, the event took place but could not be carried out conclusively to the end.
And that's exactly the kind of situation that leads to problems in the real world. The loss
of, for example, 2 euros for a drink is certainly bearable, and you will not hire a lawyer
to assert your rights. However, if it comes to more extensive contracts that have nested,

---

[11] Cf. Hopf/Picot (2018), p. 114.
[12] Cf. Fertig/Schütz (2019), p. 271.

interdependent if-then relationships, and if there is much more money involved, then the limitations of smart contracts become apparent. It is not always clear how to interpret a will legally. There are discretionary powers, even in case law. But a programmable contract requires the if-then sequence.

So not everything is "smart" (in the sense of clever) in the real world, even if it is called that.

An extended dimension is the jurisdiction, because a blockchain network will be cross-border, especially if it is a public network. It is not clear which law from which country is to be applied.

This also requires a lot of research work in this context in order to find good solutions. In the meantime, however, everyone who uses smart contracts should strive to define the conditions as precisely as possible. This cannot be left to the software developers. Rather, those responsible must define the necessary If-Then sequence together with experienced lawyers.

In order for a smart contract to be "smart" in the sense of clever, it must be able to access data and events that are outside its own blockchain world. If "only" data from the real world is to be implemented in its own blockchain cosmos, various oracle services that act as an interface between a data source from the real world and the corresponding smart contract are to be used.[13] These services transmit data from a data source to the corresponding contract.

With an Oracle program, the transactions can be brought together as the most important part of the blockchain in such a way that changes remain traceable and transparent. This is to be illustrated by way of an example with a money transfer:

You would like to exchange some euros for a digital currency. Put simply, the following happens: The euro amount is debited from your bank account and credited to your wallet address as a digital purse. In other words, something is deducted from your bank account and added to your digital account (credited). Technically speaking, these are two processes that are linked together and therefore dependent on each other. Two possibilities are conceivable:

- The money is debited from your account but not credited to your wallet, which in fact means a loss of the debited amount, i.e. a destruction of the money.
- No money is debited from your account, but it is credited to your wallet, which in fact means an increase in the money (double spending is now possible).

Oracles support blockchain databases in data maintenance in order to trace corresponding transactions. In this example, this would be to show where the money came from or where it went. Oracles as object-oriented databases allow free and arbitrary structuring of data. They help in the evaluation of conditions that cannot or can only be

---

[13] Cf. Fertig/Schütz (2019), p. 357.

insufficiently expressed via smart contracts. (Please enter the footnote number and cf. Mittelbach/Fischlin (2021), p. 338 ff)

Oracle services provide their services on-chain and off-chain, with on-chain storage meaning that the requested data is stored in the corresponding blockchain. In the off-chain variant, the data is stored on a website set up for this purpose, and this information is sent to the smart contract via (another) transaction, which as the "initiator" has ensured that there is a data request including a transaction.[14]

The awareness must be created of the logic according to which smart contracts are to be structured content-wise. In addition, the attention must also include the technical integration. Chap. 4 is dedicated to process analysis. Detailed process knowledge makes it possible to describe and stimulate the "if-then-sequence" as a smart contract content-wise and to suggest how this should be implemented technically.

One challenge today (as of 2022) is to link smart contracts that are processed across different blockchains. What is relatively easy in the analog world, namely linking one contract to another, causes great difficulties in the digital world. In the analog world, reference can be made to the respective contracts and supplemented by clauses. Technically, this is also possible in the digital world without any problems. However, the digital world currently still comes up against boundaries; different blockchain protocols and -systems cannot communicate with each other barrier-free.

This linking of the systems or interoperability is currently not satisfactorily solved. However, various companies are working on this problem worldwide, and it is therefore "only" a matter of time before the first market-ready solutions appear on the market.

3.3 Digital Currencies and Other Examples of Applications

The "mother" of all blockchain applications, as already described in Sect. 1.1, is the application of Bitcoin. Even though it seems like everything has already been said about the history of Bitcoin, here is a brief summary:[15]

In order to be able to properly place blockchain technology, it is helpful to take a step back to the year 2008 when the financial crisis, which originated in the United States, spread like wildfire across the globe. Economists often compare it to the "Great Depression" at the end of the 1920s.[16] Before the great financial crisis of 2008, there was a flood of irresponsible mortgage loans in the US, accompanied by a failure of financial supervision and regulation.

To understand this extent, one more step back in time is required. The 9/11 attacks triggered a flood of money of unprecedented proportions to overcome both the wounds

---

[14] Cf. Berghoff et al. (2019), p. 31.

[15] Adam (2019).

[16] Sinn (2010), p. 19.

of the attacks and the wounds of the bursting of the dotcom bubble. America was in a recession, and then-chairman of the Federal Reserve Alan Greenspan decided to pursue a policy of cheap money out of fear of deflation. This cheap money allowed low-income earners (sub-prime borrowers) to purchase homeownership. The credit-based demand for housing was unrestrained in the years following the attacks, and house prices soared.[17] Through clever bundling of credits (so-called pooling), new and barely regulated investment products were created on the part of banks. These products were often not understood by the actual bank managers because algorithms-driven calculations have made new standards and methods possible.[18] Roughly speaking, there were two assumptions that drove the market:

On the one hand, the assumption that real estate is a stable value investment that may stagnate in value at worst, but otherwise will rise, as well as on the other hand, the then-low interest rate policy of the Federal Reserve, which led yield-seeking investors into riskier segments.

Banks pooled supply and demand with new investment forms, of which most participants did not know exactly how this investment is composed in detail.

It can be said in retrospect that regulatory and supervisory authorities have failed and thus contributed to the collapse of the financial market.[19] This phenomenon, originally limited to the USA, could spread around the world like a domino effect as the new financial products were sold to both institutional and wealthy individual investors worldwide. And thus the American sub-prime crisis is still held responsible for the ongoing financial crisis, which in autumn 2008 drove the investment bank Lehman Brothers into insolvency as a prominent victim.

At this point, central banks around the world had to intervene and take action in the markets to prevent a complete collapse. However, the central banks were unable to prevent previously privatized profits from now having to be borne by taxpayers in times of crisis in order to keep the banking system and thus the economy alive.

It is in this economic environment that the blockchain technology first came onto the market. The intention of the inventor Satoshi Nakamoto is and was to do away with intermediaries who slow down, make expensive and inefficient a system like the banking system. With a so-called peer-to-peer network, these intermediaries are no longer needed, payments can be processed digitally just as easily as with the traditional currencies such as dollars, euros, yen, etc. (so-called fiat money).[20]

As a digital currency, Bitcoin does not exist physically, but only as an entry in the virtual ledger, which anyone interested can view. This ledger, as a blockchain database, logs every transaction.

---

[17] Mallaby 2016, pp. 596, 597.

[18] Bloss/Ernst/Häcker/Eil, 2009, e.g. pp. 69 et seq.

[19] Bloss/Ernst/Häcker/Eil, 2009, e.g. p. 69 ff.

[20] Nakamoto, 2008, p. 1.

To whet your imagination, a small selection of possible fields of application is presented below, where it is possible to introduce corresponding projects.

▶   Even if your idea is already in one of the presented application areas, this is no
      reason why you should not pursue your idea. There is virtually nothing that
      does not exist. It is only up to you how you improve an existing (also block-
      chain-based) solution. Use these examples as inspiration!

Fig. 3.2 gives a rough overview of topic areas, some of which are described in more detail.

Banking

"Know your customer" is the required identification check when entering into business relationships and is considered the most important process for preventing money laundering. This means that customers must go through the process from scratch each time they enter into a business relationship with a KYC obliged entity such as a bank, fund company or insurance company, etc. This includes disclosing their identity to the institution. This is time-consuming and costly for both sides. According to the position paper of the Association of German Banks, inconsistent requirements for verifiability and reuse of the data to be collected prevent this process from being carried out digitally or cross-border.[21] Blockchains could provide a remedy, as the data stored on them is not manipulable.

The global anti-money laundering directive is heading in the same direction. It is about the prevention and combating of financial crime, in particular money laundering and terrorist financing.

Financial Services

The Bitcoin blockchain is the idea that digital money can be given directly from one user to the next, just like cash. But this doesn't exhaust the processes in banking. Companies along the financial services value chain are currently investigating how this technology can be used for the processes that typically arise. While some may dream of eliminating the intermediary bank (or other financial service provider), the market is still a long way from this. This may be due to missing rules and legal requirements as much as simply the fact that some companies cannot take on this challenge of digitalization. Whether these companies are future-proof is not the subject of this discussion.

---

[21] Position paper of the Association of German Banks (2019): Verification of corporate customers in the EU internal market.

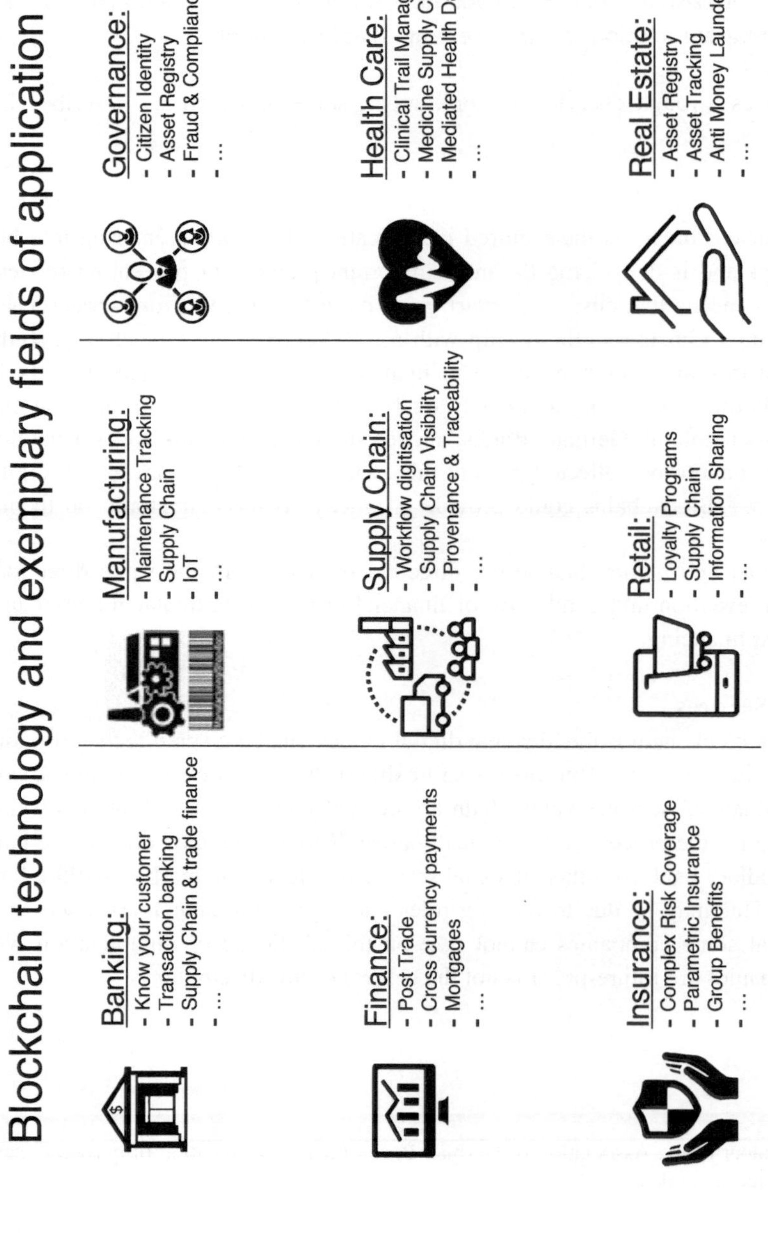

Blockchain technology and exemplary fields of application

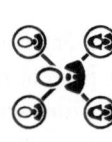

Banking:
- Know your customer
- Transaction banking
- Supply Chain & trade finance
- ...

Manufacturing:
- Maintenance Tracking
- Supply Chain
- IoT
- ...

Governance:
- Citizen Identity
- Asset Registry
- Fraud & Compliance
- ...

Finance:
- Post Trade
- Cross currency payments
- Mortgages
- ...

Supply Chain:
- Workflow digitisation
- Supply Chain Visibility
- Provenance & Traceability
- ...

Health Care:
- Clinical Trail Management
- Medicine Supply Chain
- Mediated Health Data Exchange
- ...

Insurance:
- Complex Risk Coverage
- Parametric Insurance
- Group Benefits
- ...

Retail:
- Loyalty Programs
- Supply Chain
- Information Sharing
- ...

Real Estate:
- Asset Registry
- Asset Tracking
- Anti Money Laundering
- ...

Fig. 3.2 Possible applications of blockchain

The Fraunhofer Institute points out in its study that "today's payment processes still involve several intermediaries, e.g. banks as well as clearing and settlement offices."[22] Many processes that arise are not processed in real time, i.e. immediately, but bundled and time-delayed. This can mean that a special process is only carried out once a day. This can lead to unnecessary delays, which were not minimizable so far for technical reasons. The often non- or only semi-automated processes therefore take up a lot of time—also working time. The time-saving potential therefore appears to be large and lucrative.

The blockchain technology can be used to irreversibly and simply compare data during financial trading. In addition, it helps to increase the effectiveness and speed of the comparison (since this is done in real time), and this helps to increase the security level of transactions between the parties involved in the purchase and sale and their banks.

Prediction markets that describe future trading are now being emulated and extended using blockchain-based solutions (e.g. companies like "Augur" or "Gnosis"). Arbitration opportunities are used by companies (e.g. Kraken) to exploit the high volatility of cryptocurrencies and tokens. Crowdfunding can also be mapped using blockchain to enable mass-based risk financing (e.g. NeuFund).

Supply Chain

Another segment that seems almost predestined for blockchain applications is supply chain management. Along a supply chain, huge amounts of data and opportunities arise. This data is easy to manipulate because many processes are paper-based and manual. Missing information about, for example, the transport history testifies to a lack of transparency. This can have negative consequences:

- unexpected cost increases,
- slowed down own production processes due to disruptions in the supply chain,
- lack of traceability,
- loss of quality.

Blockchain solutions make it possible to trace in real time where products are located within the supply chain. Research is also being conducted into how audit procedures along the supply chain can be made more sustainable and reliable through the use of a blockchain.

Real Estate

There are also numerous processes in this industry that can be improved using a blockchain solution. The classic example is registration entries, e.g. in the land registry office. By using the blockchain, ownership rights can be transferred directly between the seller

---

[22]Prinz et al. (2017), p. 50.

and buyer. A notary is no longer required as an intermediary. The Swedish state is currently running a test version with the company CromeaWay to gain experience in these procedures. The author of this book has written a white paper on the transfer of property rights, taking into account the special features of the German land register.[23]

Healthcare

A blockchain solution enables personal data to be entered into and stored securely in a patient record. This allows patient data to be securely made available to or transferred to various doctors or health insurers.

The monitoring of the supply chain for medication offers similar opportunities as already listed under "Supply Chain".

In addition, it is being discussed to use blockchain solutions in the cultivation of medical cannabis. The first projects in this regard are located in Silicon Valley, and at least theoretical treatises are also available in Germany.

Furthermore, it is possible to store and document drug research via a blockchain in order to promote, for example, a approval process by proving that no manipulation of the results has taken place.

Energy

In the energy sector, too, there are many areas that can be checked. It is assumed that the energy system can be better organised and made more transparent for all participants by using blockchain. New business models are emerging. It is conceivable that private producers sell their solar power directly to a user without previously feeding the generated surplus into the public network of an energy supplier before the end user can use it for his/her purposes. On the website of the *Energy Web Foundation* you can get an overview of the various possible applications that are currently being considered (cf. DApp): from the reorganisation of an energy exchange to Charging-as-a-Service (interesting for electromobility) to consumer-oriented billing models, fields of application and corresponding companies are presented that are developing mass-compatible solutions.[24]

Media Industry

With the help of a blockchain, the existing chaos of rights and licenses in the music industry could be resolved. This market is highly fragmented, and the artist does not always receive the money due to him. Blockchain can secure rights here.

The sale of publications or other usage rights in chapters can also be mapped using a blockchain solution.

---

[23] Adam (2017).
[24] Energy Web Foundation (2019), http://energyweb.org/about/what-we-do/; accessed on 28.09.2019.

Public Sector

- *Examination of certificates:* School and academic performance is securely stored by the issuing institutions and can be forwarded to other institutions and/or employers.
- *Voting systems:* Elections remain anonymous and are protected from counting errors. Voters can vote from home, which can increase voter turnout.

3.4   Security Aspects

The properties of the blockchain have been illuminated in the previous subchapters, so there will be no repetition here. But it is also clear that blockchains are also subject to attacks. The properties of the blockchain (no single point of failure, because distributed network and P2P; cryptographic elements such as asymmetric key pairs, hash functions, high data quality, tamper-proof, anonymity or pseudonymity) make it difficult for hackers to attack and "crack" blockchains.

For reminder: Once a transaction has already been sealed in a block and added to the blockchain, it is almost impossible to change it. Not only would it be necessary to retrace the hash block and make a change to the transaction data contained therein, but this would have to take place simultaneously with at least 51% of the copies held on different nodes. Therefore, it is practically impossible to "hack" a blockchain.

However, an attack can also take place from "within", and then one speaks of a 51% attack. The larger a network is, the more difficult it is for attackers to achieve the necessary majority. If they succeed, the hackers can decide to block transactions, exclude them or change the order of the transactions, which would cause overall massive damage.

The larger a network is, the more difficult this hostile takeover will be. In addition, the history of stored transactions makes it more difficult, because the blocks with the transactions made are linked to each other via a hash value. And it applies: The more confirmations a block has, the more difficult and costly it will be to try to change it.[25]

In this context, it is often warned that mining pools could decide to carry out such an attack. This may make sense if you want to destroy the Bitcoin blockchain, for example, for ideological reasons. But as long as money can be earned in the form of Bitcoins through mining, such a hostile takeover would destroy confidence in this digital currency. Then the coins held by the attackers would also be worthless.

The probability of such an attack is very low. So far, the Bitcoin blockchain has not been "hacked" maliciously. The successful attacks that are reported relate to cryptocurrency

---

[25]Cointelegraph: Exceptions confirm the rule: In May 2019 there was such an attack on the Bitcoin Blockchain. This attack was considered a good intervention to "prevent" coins that were created by a previously performed update from being available to unauthorized persons.

platforms. The effort to crack such a platform is "easier" compared to the actual block-chain, because only one node has to overcome the firewall.[26]

3.5   Blockchain "Value"

This technology also has to answer the question of whether it provides added value for companies and society. Blockchain technology is one of the technologies that will greatly influence or influence the change in our society. The way contracts are orches-trated, how collaboration is redefined and how roles change within processes will be sig-nificantly influenced by this technology. No industry or institution can exempt itself from this.

The advantages of the blockchain are often described as confidence-building, open, cross-border, fast, safe, tamper-proof, real-time processing. The advantages of the net-work are to be seen as greater than those of a centrally oriented structure. If the latter falls, the organization also falls. In decentralized networks, however, it is irrelevant whether a node fails. The network, which refers to all computers involved, ensures that all computers always have an up-to-date and synchronized version. If a node is switched on again, it is updated according to the current state. There is no central instance in such networks that makes decisions. This distributed (democratic) approach changes the way processes are handled.

Blockchains are based on their network, and the network can become very strong and robust. This allows companies and institutions to develop network structures that give them a new dimension of process structuring.

Companies like Uber and Airbnb show the power that networks develop, although it should be noted that these companies in particular have a "the-winner-takes-it-all" understanding. Real collaboration can, however, arise anew and with a reduction in asymmetries through the use of blockchain. Exchange and communication take place in real time. These blockchain networks are not limited locally, but global and secure. This makes a (blockchain) network a value in itself.

When new network structures arise, the direct exchange of data (whether transactions or other information) becomes easier to carry out than before, disrupting existing busi-ness models. The roles of currently involved intermediaries such as notaries, banks, state energy providers, etc. will change fundamentally.

Smart contracts as program code support the automation of processes and thus con-tribute to the shortened transaction processing, since the predefined software code ideally does not require human intervention.

---

[26] Mochizuki/Warnock: Mt Gox: At the 2014 hacked platform, the attackers were able to steal 850,000 Bitcoins.

These advantages are offset by disadvantages. Technically problematic is the current transaction rate (through-put), which is not yet (!) up to the processing of massive data. The proof-of-work consumes too much energy, which is of increasing importance in the sharpening climate debate. The other consensus models listed show the desire to overcome this deficiency.

The storage capacity of a classical blockchain can grow infinitely—if it were technically possible.

But perhaps the biggest hurdle of this technology is the transformation from a central to a distributed approach. The existing understanding of processes and responsibility is still strongly shaped by hierarchy. This necessary "mind-shift" is the real challenge, as it redetermines positions and responsibilities.

3.6   Conclusion

In Sect. 1.1 the phases are described that the blockchain technology has gone through. If you base the phases on and expand the approach, you will see four categories in Fig. 3.3 that cause different requirements.

The initial goal relates to services and solutions to increase efficiency and effectiveness. So-called geeks and nerds, that is, people highly familiar with technology, develop niche knowledge and work on these not yet popular topics. It is not mainstream yet. Mathematical skills are in the foreground, for example, to further develop cryptographic

The road so far ...

	Cryptography & Digital Currency	Satoshi Nakamoto & Ethereum	New Industry
Output & Outcome	Services/Solutions; Efficiency & Effectiveness Improvement	Elimination of intermediaries, process optimisation	Blockchain 3.0; solutions for business, public sector and society
Engagement	Niche knowledge; Geeks & Nerds	Peer-to-peer approach	All stakeholders; security & trustworthiness
Skills	Mathematical skills; central concepts for integrity and authenticity	IT management, programming	Process understanding; management skills IT understanding
Focus	Engineering, technology and encryption	Financial sector and smart contracts	Hybrid approach; for business and society; technology as enabler

Fig. 3.3 Requirements according to the phases

concepts. This focuses on the technology with the technology to be used and new encryption and transmission concepts.

From this, the blockchain ecosystem develops under Satoshi Nakamoto. Intermediaries are now obsolete, processes are optimized. The network with the peer-to-peer approach becomes the driver of development. IT management and programming are the skills required in addition to a mathematical understanding. First promising concepts are realized for the financial sector, also under the embedding of smart codes. This is the time of the "early adaptor".

This technology is no longer in the niche, and from the experiences of the Early Adaptor, more and more market participants are recognizing the potential for their own (business) processes. All stakeholders must be involved so that the newly developed approaches can meet with broad approval. In the extended use of this technology, the skill set of those responsible must also be extended. In addition to IT understanding and general management skills, process understanding is of elementary importance. The focus expands to mass-usable applications for all areas of society. The technology is understood as an "enabler", as an "enabler".

Overall, the blockchain community is on the way to transforming itself into a new industry.

References

Adam K (2017) Project Hurricane – or how to implement Blockchain Technology in German Real Estate Transactions. Diskussionspapier

Adam K (2019) Blockchain – die etwas andere Datenbank; Sonderband 04/2019 „Die Modellierung des Zweifels; Schlüsselideen und -konzepte zur Modellierung von Unsicherheiten". Zeitschrift für digitale Geisteswissenschaften (ZfdG), Wolfenbüttel

Antonopoulos AM (2017) Mastering bitcoin, programming the open blockchain, 2. Aufl. O Reilly Media, Sebastol

Bankenverband (2019) Positionspapier: Privatkundenverifizierungen im EU-Binnenmarkt. Berlin

Berghoff C, Gebhardt U, Lochter M, Maßberg S et al (2019) Blockchain sicher gestalten. Konzepte, Anforderungen, Bewertungen. Bundesamt für Sicherheit in der Informationstechnik (Hrsg) Bundesamt für Sicherheit in der Informationstechnik, Bonn

Bloss M, Ernst D, Häcker J, Eil N (2009) Von der Subprime-Krisis zur Finanzkrise. Oldenbourg, München

Buterin V (2013) Ethereum white paper: a next generation smart contract & decentralized application platform. http://blockchainlab.com/pdf/Ethereum_white_paper-a_next_generation_smart_contract_and_decentralized_application_platform-vitalik-buterin.pdf. Accessed: 22 June 2019

Cointelegraph (2019). https://de.cointelegraph.com/news/two-miners-purportedly-execute-51-attack-on-bitcoin-cash-blockchain. Accessed: 22 June 2019

Energy Web Foundation (2019). http://energyweb.org/about/what-we-do/. Zugegriffen am 28.09.2019

Fertig T, Schütz A (2019) Blockchain für Entwickler, Grundlagen, Programmierung, Anwendung. Rheinwerk, Bonn

Hopf S, Picot A (2018) Revolutioniert Blockchain-Technologie das Management von Eigentumsrechten und Transaktionskosten? In: Redlich T et al (Hrsg) Interdisziplinäre Perspektiven zur Zukunft der Wertschöpfung. Springer, Wiesbaden

Kersken S (2019) IT-Handbuch für Fachinformatiker. Der Ausbildungsbegleiter, 9., erw. Aufl. Rheinwerk, Bonn

Mallaby S (2016) The man who knew: the life & times of Alan Greenspan. Bloomsbury, London/New York

Mittelbach, Arno/Fischlin, Marc (2021): The Theory of Hash Functions and Random Oracles, An Approach to Modern Cryptography. Cham, Springer Nature

Mochizuki T, Warnock E (2014) So lief die spektakuläre Pleite der Bitcoin-Börse ab. Wall Street Journal. https://www.welt.de/wall-street-journal/article129565422/So-lief-die-spektakulaere-Pleite-der-Bitcoin-Boerse-ab.html. Accessed: 20 Sept 2019

Nakamoto S (2008) Bitcoin: a peer-to-peer electronic cash system. https://nakamotoinstitute.org/bitcoin/.pdf. Accessed: 22 June 2019

Prinz W et al (2017) Blockchain und Smart-Contracts; Technologien, Forschungsfragen und Anwendungen. Positionspapier, Fraunhofer

Sinn H-W (2010) Kasino Kapitalismus. Wie es zur Finanzkrise kam. Ullstein Buchverlage GmbH, Berlin,

Standard for Efficient Cryptography. http://www.secg.org. Accessed: 16 Oct 2019

Szabo N (1997) Formalizing and securing relationships on public networks, pdf. https://nakamotoinstitute.org/formalizing-securing-relationships/. Accessed: 31 Aug 2019

"Tools of Action" (Process Analysis)

4

Abstract

This very comprehensive chapter explains how you can approach this technology. Various management techniques are described and it is shown how the use of these methods helps to illuminate your ideas and processes from different perspectives. The workshop character is in the foreground in order to discuss in interdisciplinary teams and to achieve sustainable solutions.

When starting this chapter, I would like to ask you to do the following: First of all, forget about blockchain technology and blockchain engineering!

Much more important is to find out which processes exist and where these existing processes are slow and inefficient. Only later does the question arise as to whether these processes and their sub-processes require a blockchain-based solution—or whether existing solutions "only" need to be revised. Of course, the connection between the tools and the blockchain technology will be established in the individual steps listed here. However, tools are in the foreground.

▶ Please do not limit yourself by thinking of the solution from the outset. So if you approach your process analysis with the intention of mapping it to a blockchain solution, it could happen that other, possibly even better solutions for this application do not occur to you.

Your blockchain solutions must also create value, otherwise they will only trigger a lot of effort and high error rates without providing any added value. From a business point of view, this means wasting resources of any kind (financial, personnel and asset deployment). At the end of this chapter, you should be able to specify the requirements for

K. Adam, *Blockchain Technology for Business Processes*,
https://doi.org/10.1007/978-3-662-65818-5_4

improving processes. And if it makes sense to map this process to a blockchain solution, you should be able to formulate this precisely so that your developers understand what you want and need. Your developers do not have to be in-house developers. The market offers many possibilities for "modular" purchase, i.e. you can get the tools you need tailored to your needs on the market. But, and this is important, you must be able to define very precisely what you need and how.

▶   For the sake of readability, the male form is chosen in the text, of course the information always refers to members of both sexes.

4.1   Preparation

First, it makes sense to take a closer look at the stakeholders who are involved in your company processes. In order to make the subsequent steps comprehensible, I allow myself to show you the example from my lesson in which I describe my university as an inquiring entrepreneur.

Furthermore, I recommend that you agree within the team who will hold the group and thus the moderation and discussion leadership before the start. It is possible to define the project manager as the group leader. However, it is not necessarily the case that the project manager also manages the moderation and discussion—this must be discussed in advance.

Also, agree in advance who will document the workshop and who will create the report at the end. The better the documentation, the more you involve the participants, and the findings and results have a wide basis of agreement.

To carry out the following steps actively and with plenty of room for discussion, it is recommended to have the following (typical) workshop material available in addition to a suitable room:

- Moderation cards in different colors, shapes and sizes,
- Pens in different thicknesses and colors,
- Tesa film,
- Post-it or small index cards,
- Flip chart paper,
- Scissors,
- Magnets,
- DIN A4 paper.

Before you start, please take a look at your current status quo using the so-called gap analysis.

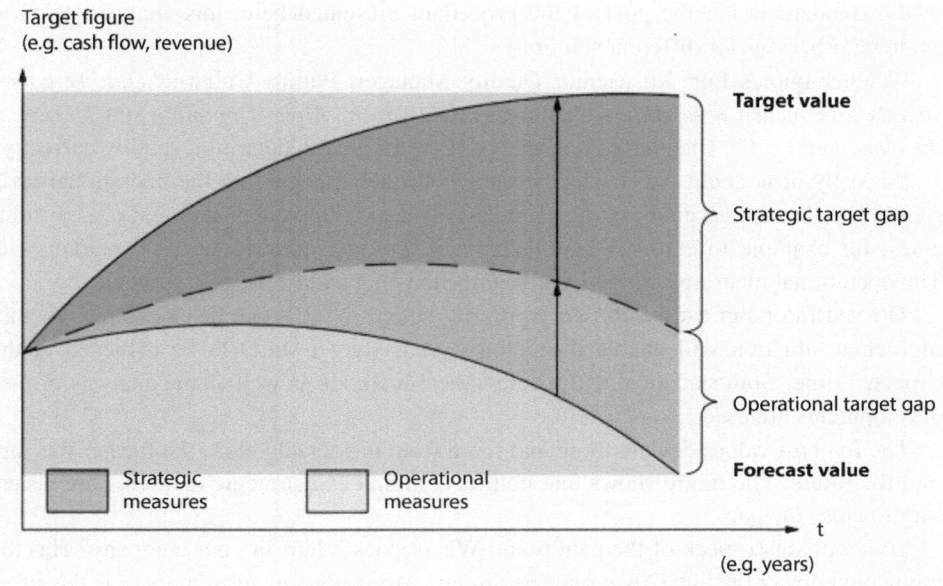

Fig. 4.1 Gap analysis. (From Jean-Paul Thommen, Ann-Kristin Achleitner, Dirk Ulrich Gilbert et al. 2017; with kind permission from © Springer Fachmedien Wiesbaden GmbH 2017. All Rights Reserved)

4.2   Gap Analysis (Gap Analysis)[1]

This classic element from strategic management theory helps companies determine where they stand.[2] It is therefore a status analysis. In addition, it can be determined whether one's own interpretation of the current status is going in the right direction. If you understand your current status (and document it), it is easier to formulate the new goals to be achieved. With this inventory analysis, activities can be defined in order to achieve the goal and close the gap found (cf. Fig. 4.1).

Relevant target variables are, for example, sales, profit, contribution margin or value added. These target variables can relate to departments, product groups, business units or the entire company.

In addition, it is recommended to deal intensively with the ISO standard 9001 in order to understand the procedure and evaluation described in the standard. Even if the standard requires a management system in the company and in addition certification according

---

[1] Koubek (2015) ISO 9001, Version 2015.

[2] Thommen et al. (2017), p. 537.

to the standard is not the goal of the procedure presented here, this analysis helps to examine processes for different solutions.

Weidner quotes Len Xu (Senior Quality Manager, Philips China): *"Quality corresponds to a measurable state which corresponds to the highest possible satisfaction of the customer. [...]"*[3] The goal of this analysis is to be able to determine quality correctly.

Basically, it is about two levels, because both the strategic and the operational level should be taken into account. Strategic measures deal with the development of new strategies, for example to consider new techniques, product innovations, new markets etc. The operational measures are derived from the existing and/or the new strategies.[4]

The starting point can be chosen arbitrarily at first. Now it can be calculated whether the actual situation will enable the defined goals (target state) to be achieved at the expected time. Both statistical and mathematical methods as well as the analysis of past developments are used.

The forecast values deviate more and more from the actual values the further they are into the future. The figure shows that both operational and strategic measures are necessary to close the gap.

You could also speak of the pain point: Where does it hurt in your company? Has the innovation power declined? Are sales declining? Are you losing customers or is the effort to win new customers much more expensive? Are your products and the processes to produce them still state of the art? Which processes are more expensive and slower than necessary? Which technology needs to be used to better manage the processes?

You will have your own pain point that points to gaps. The gap analysis may be a simpler tool (unless you are aiming directly at ISO certification), but it is widely used in practice because with this analysis

- it is possible to capture the factors that have a shaping effect on future development and
- it shows the consequences if no (necessary) countermeasures are taken (sales decrease, customer base decreases, etc.) and
- it activates the search for suitable strategies to close the discovered target gap or at least significantly reduce it.

▶   Think about which gaps you see in your company as a (management) team. Which processes are behind these gaps? If, for example, purchasing causes rising costs, then there is a problem in procurement. Which of the many processes in purchasing alone causes higher costs than expected, etc.?

---

[3] Weidner (2017), p. 31.
[4] Dillerup/Stoi (2013), p. 301.

Table 4.1 Comparison Blockchain vs. traditional database

Blockchain	Properties	Traditional database
Strictly additive, only additive	Progression in usage (operations)	Insert, delete, change, supplement, rewrite, etc.
Yes	Redundant	Yes
Yes	High availability	Yes
Block-wise, e.g. via so-called Proof of Work	Consensus	Series, replica (duplicate)
Always necessary	Signature	Can be manually inserted
Always	Data validation	Manually
Smart Contracts	Business logic	Storage procedures
Distributed ledger logic (Ledger)	Primary Use	Generic Approach

4.3 Options

First of all, blockchain technology can be explained in one sentence: It is a database, although a special one. The main differences between a traditional and a blockchain database are listed in Table 4.1.

The individual characteristics of a blockchain are explained in Chap. 1. It can be seen from the table that the peculiarity of a blockchain as a tool actually lies in the unchange-ability of the data entered and that the data can "only" be appended. The data entered cannot be "simply" deleted or overwritten. In the event of a change, a new transaction must be triggered, which is also to be stored on the blockchain in order to trace the history.

If you look a little deeper into the consensus mechanism, you will also see an almost endlessly appearing (possibly also confusing) variety. The decision of when which consensus applies to which type of blockchain cannot be fixed. So a public blockchain does not necessarily have to go hand in hand with the proof-of-work consensus as with the Bitcoin blockchain. The variations that open up depend on the business processes for which a blockchain-based solution is to be designed.

To check business processes for "blockchain suitability", at least 169 different approaches are available based on the assumption of three blockchain types with 13 consensus mechanisms each. As nice as the choice options are, the decision matrix becomes complex. In order to be able to make decisions from the wealth of offers, the "tool box" is opened below, with the help of which it can be filtered out whether a process can be mapped profitably via a blockchain solution—or not.

The process begins with the so-called stakeholder analysis to consider which groups of people have what influence on the company.

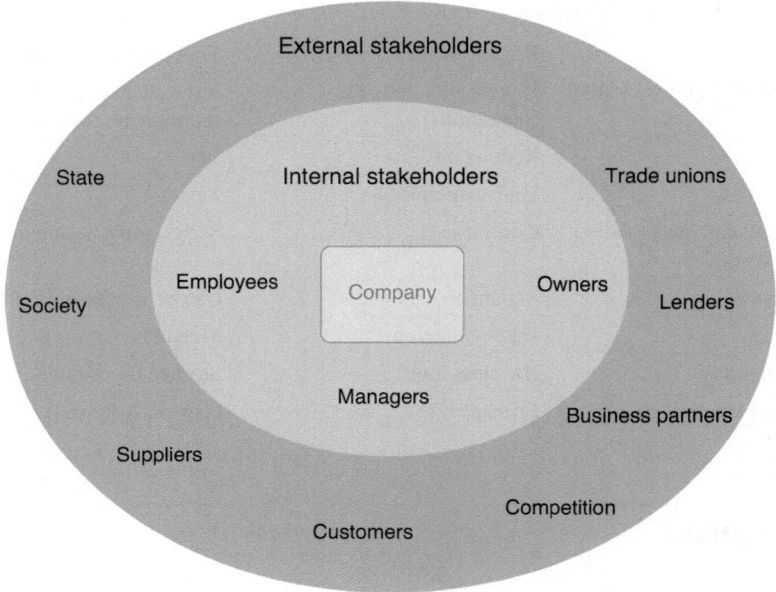

Fig. 4.2 Stakeholder groups of a company. (own representation based on Dillerup and Stoi 2013)

4.4   Stakeholder Analysis[5]

The stakeholder concept dates back to 1963 and was developed at the Stanford Research Institute (SRI) to sensitize managers to other interest groups than the shareholder (owner). In management literature, the following definition has established itself:

▶ **Stakeholders** are people, groups or organizations that are related to the company and have expectations of the company.[6]

A distinction is made in stakeholder analysis between internal and external stakeholders, as these different segments have quite different requirements and expectations of a company (see Fig. 4.2). Mintzberg speaks here of influencers, that is, those who influence the organization.[7]

---

[5] Din 69901-5.

[6] Dillerup/Stoi (2013), p. 118.

[7] Mintzberg (1983), pp. 26 ff.

Both groups of claimants (internal and external) must be included in the company's strategy. In addition, management should be aware of how the company's strategy should look like towards these groups of claimants, whether and when it is a question of action or reaction.

With regard to projects (and the examination of whether an internal company idea can be a worthwhile project for a blockchain solution), it is therefore important to identify and assess all relevant actors and to what extent these stakeholders could influence the company and the project.

In principle, one can approach the classification of stakeholders through the following questions:[8]

- Who are the current stakeholders—external or and internal?
- Who are possible stakeholders?
- What influence does each stakeholder have on the company?
- What influence does the company have on each stakeholder?
- Who are the stakeholders within the company?
- How important are the individual stakeholders for the company's success?
- What are the factors and their variables that have an impact on the company?
- How are these variables and their impact on the company or other stakeholders measured? Can these variables be measured at all?
- How are the stakeholders observed?

Let's start with the identification of the stakeholders from the above areas. Methodologically, we start with brainstorming, a creativity technique that allows a large number of first ideas and solutions to be generated within a very short time. However, this only works if some conditions are met. First of all, it is about quantity and only in the second step about quality. This creates a separation between idea finding and idea evaluation.

The ideal group size is between 4 and 8 people for both brainstorming and the following tools. Larger groups should be divided into subgroups.

The group leader must make sure that despite the quantity, the topic is not deviated from.

▶   **Tip** Distribute Post-its, small cards or similar to your group participants, on which key points can be noted, and let each person write down all stakeholders that come to mind for the individual group participant. As always with brainstorming—there is no limit to thinking, no evaluation of what has been recorded.

---

[8] Pastoski (2004), pp. 5,6.

After a maximum of 10 min, all key points are first placed on a stack, and duplicates are sorted out.

It is important for the group leader at this point not to allow any assessment of the individual key points, otherwise the group will commit too early and thus ultimately limit itself.

Now that the stakeholders have been identified, it is good to evaluate the found stakeholders according to influence, power, conflict potential and possibly worldview, attitude, wishes, hopes, etc. Take your time in the group to discuss so that the team has a common understanding of the importance of the stakeholders for the company or the process to be improved.

▶ Stakeholders are potential nodes and/or participants within your blockchain solution!

The collected and ranked information can now be transferred to the stakeholder map (cf. Fig. 4.3).

It is recommended to position the found stakeholders on an axis according to their influence at first, in order to only apply the other axis/dimension in a second step.

This stakeholder analysis is exemplarily shown for the University of Applied Sciences for Engineering and Economics (HTW) Berlin:

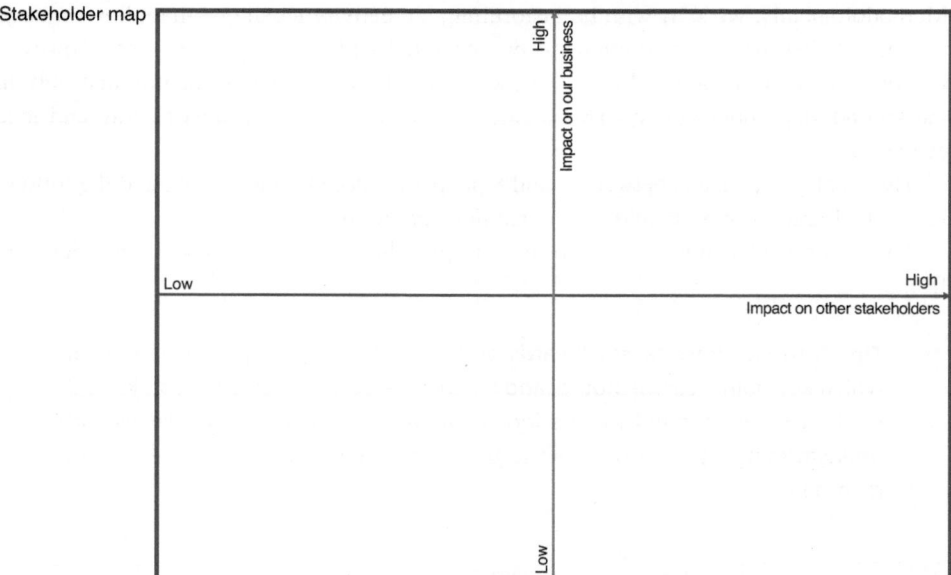

Fig. 4.3 Stakeholder map

Fig. 4.4 Example stakeholder identification

First, the stakeholders (the list can be extended arbitrarily) are identified (cf. Fig. 4.4).

The identified stakeholders are now ranked. It should be noted here that this is always a subjective assumption. It is therefore important that the group discuss this in the workshop. This can of course lead to controversy, because each participant has his/her or her own individual view. However, at the end of the discussion, the team leader must ensure that a stable compromise is reached. And, we're not talking about absolute accuracy here, but about estimates.

In accordance with the HTW example, the identified stakeholders are transferred to the stakeholder map as follows. As mentioned above, it makes sense to start with an axis. In this example, the stakeholders are sorted on the Y-axis according to their assumed influence on their own company (cf. Fig. 4.5).

Now you can move the individual stakeholders along the x-axis. In the example, it then looks as shown in the graphic (cf. Fig. 4.6).

From the "wild" heap of stakeholders resulting from the brainstorming session, a map has now emerged that provides a first indication of which of the identified stakeholders deserves special attention. In the example, it can be seen on the one hand that it is the students, on the other hand the employees at the different levels and in the different areas. In this context, employers are also of importance, as companies that appear

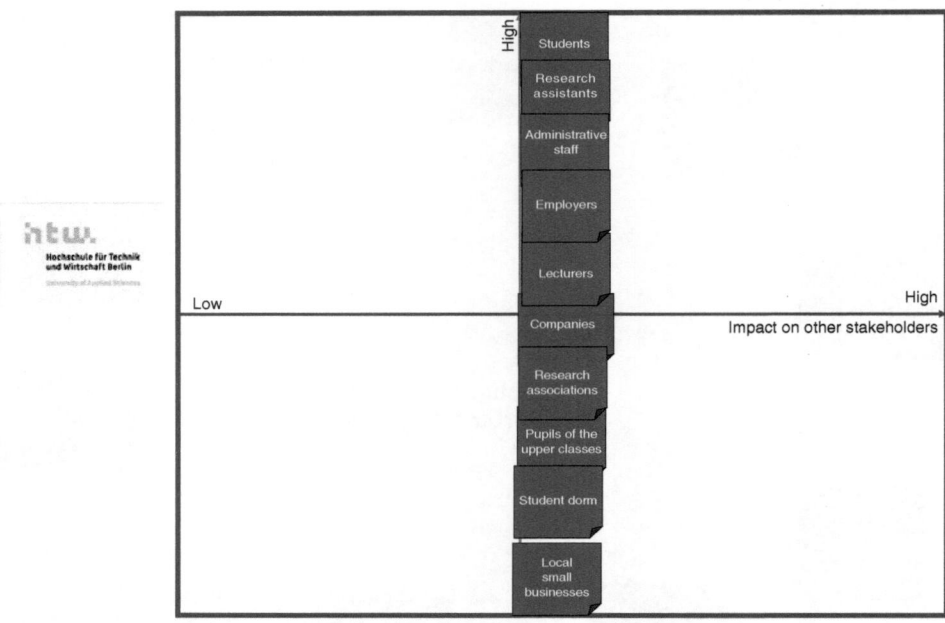

Fig. 4.5   Stakeholder map using the HTW example; Impact ranking on the Y-axis

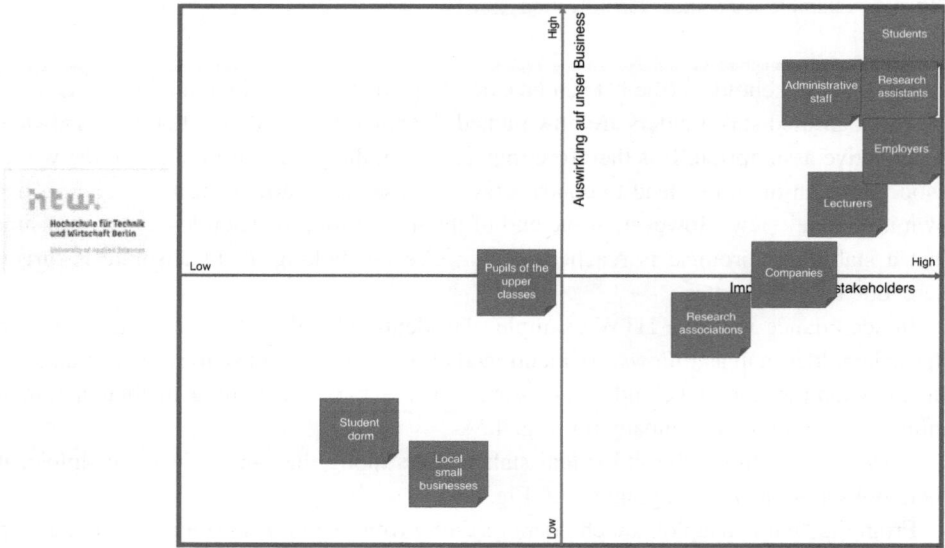

Fig. 4.6   Stakeholder map using the example of the HTW; impact ranking along the X-axis

as employers on the market expect to receive well-trained graduates from the university. For this purpose, motivated and committed employees are necessary at all levels. In summary, it can be said that the stakeholders in the upper right quadrant are significant.

▶   Identified and assessed as important stakeholders could be potential nodes of the blockchain network in an emerging blockchain solution and, depending on the consensus and protocol, actively contribute to the validation of the individual transactions to be carried out.

4.5   Products and Services Map

If the stakeholder analysis and the upper right quadrant have given you first indications of potential participants/shareholders of your blockchain solution, the analysis of your products and services will help you to get an overview of the context of blockchain solutions: Which of the products/services solve which activities with which process steps (see Fig. 4.7)?

Please proceed as follows:

- In a brainstorming session, name all of your products and services (depending on the size of the group and the total time available for the workshop, this session should last 15 to 30 min).
- Place your results in the header of the following "story map"

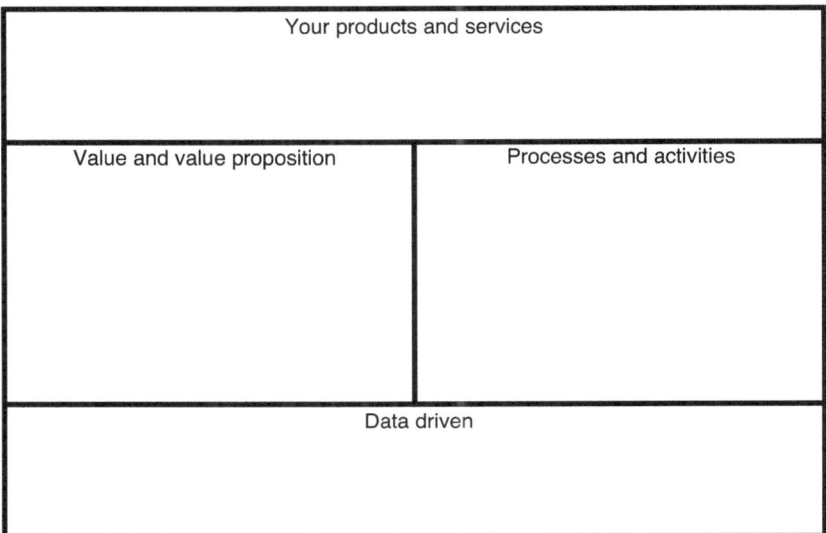

Fig. 4.7 Story-Mapping-Karte

- Identify processes/activities as well as value claims and place them accordingly on the map.
- Discuss the intermediate results as a group.
- Move the elements that are data-driven to the footer and summarize the results for all group members so that there is agreement on the status quo found so far.

The derivation of "values/performance promises" as well as "processes and activities" always leads—at least in my courses—to heated discussions among the participants. It is not always clear for the participants to identify what value is created by the existing products and services. It is about the benefit or the performance promise that is created and, according to management literature, it is possible to distinguish here between the benefit promise to the customer or to suppliers and other stakeholders. This performance promise can be considered as a declaration of intent to customers or suppliers, because it shows what a company stands for.[9] The group leader should make this performance promise conscious—for what do we stand (e.g. our department or the company), how do we create value for our customers, how do we care for our suppliers and other stakeholders, and how do we bind these groups to the company. The use and performance promises can then be relatively easily derived from the processes and activities. It should be considered here how deeply one wants to get into the process levels. It can still be "high level" at this point and thus not too deep.

Figure 4.8 shows an example of how this could look for the university (and again the list is not complete—you should only get an impression of how you could determine the positions).

In this way, "education", "events", "further education", "self-administration"; "research institutions", "research" and "food and drink" are listed as products and services. The latter was brought to my attention in particular by the students. Under "normal" circumstances, one would not list and add something like this to the product and service catalogue. But the example shows that it is the customer's (= student's) point of view. And that is important! Therefore, once again the hint not to build any evaluations and thus limitations during brainstorming.

The values and performance promises are then derived from the products and services, such as high-quality training, grades, social recognition (in the case of a degree), certificates, exchange with industry and the HTW Student Card (which includes the Mensa card function as well as the library card, the copy card and also a payment function). Overall, this list could also be completed. Do not limit yourself in terms of the number of terms during the brainstorming session, but limit yourself in time to work discipline.

The processes and activities are interwoven with their products/services as well as with their values and performance promises. In the example given, the activities listed trigger processes. As already mentioned above, you do not have to go too deep into the

---

[9] Gerzema/Lebar (2008), p. 2.

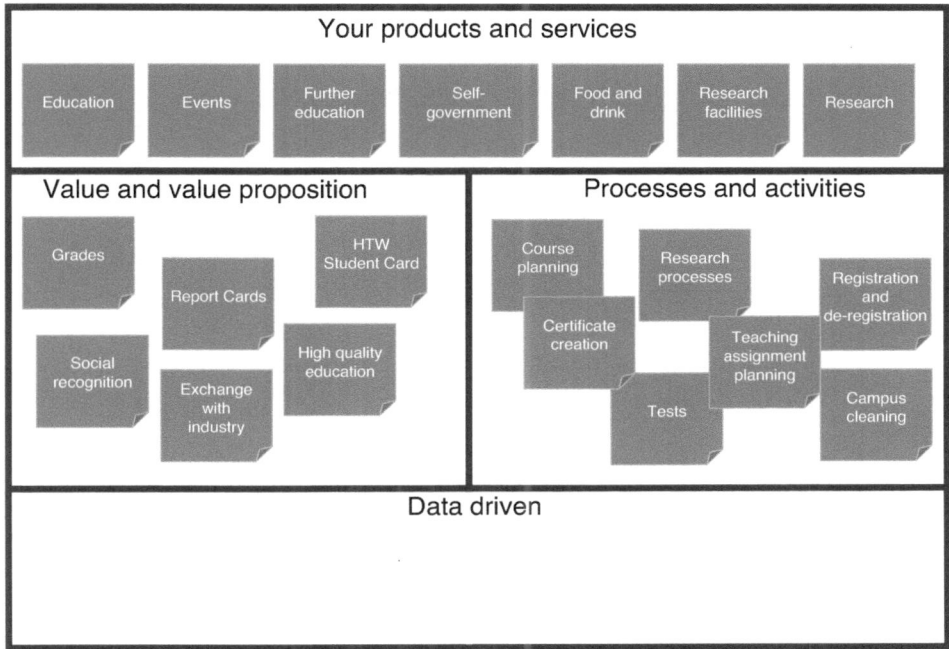

Fig. 4.8   Example Story-Mapping Card

process level here. Further discussions are desirable in order to obtain a common understanding of the results found.

The derivation of the data-driven/digitized processes and values is—depending on the discussions held—relatively easy.

▷   The knowledge of the story mapping card can show in relation to the blockchain technology which of the activities, for example, can be carried out as transactions or with the help of a smart contract. Values, on the other hand, could be used as a coin or asset in the future business model.

Value stream mapping (VSM) should be used as an extension to the findings. Value stream mapping (VSM) is a method from lean management.

This procedure systematically captures processes and can then redesign them. The literature refers to the fact that for companies responsiveness and flexibility are almost essential for survival in the market. In addition to costs and quality, the ability to quickly adjust to constantly changing market and customer requirements is also in focus.[10]

---

[10]Pfeffer (2014), p. 7.

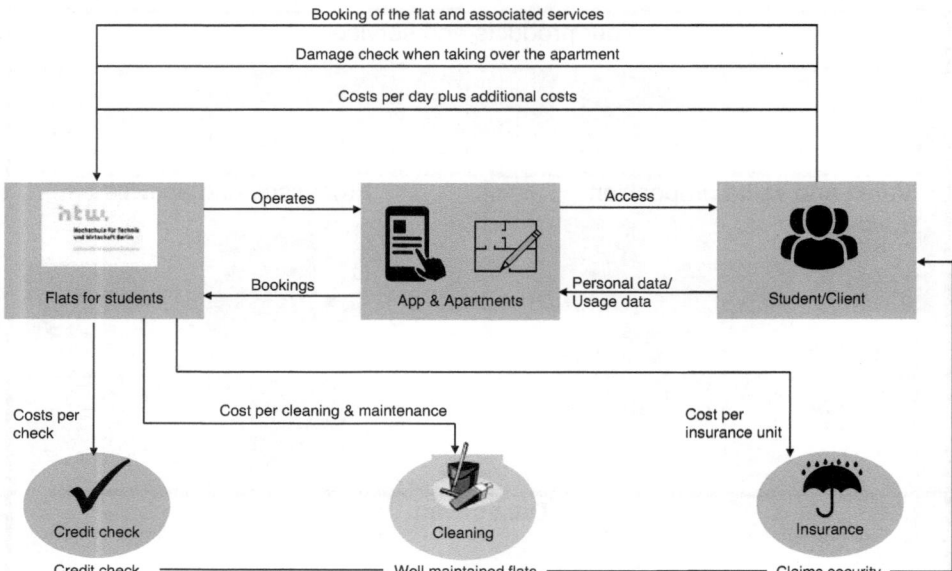

Fig. 4.9 Example Value Stream Mapping

First, the overall process should be in the foreground, before the individual processes are defined in terms of the right target specifications. This approach is very common for traditionally manufacturing companies, because the necessary process steps are defined, analyzed and optimized by reorganization.[11]

Similarly, the procedure described here is followed.

On the basis of what you have worked out so far, the task is now to find sensible combinations of stakeholder and data-driven activities.

An example illustrates this approach (cf. Fig. 4.9).

In this example, there are two participants who interact with each other via an application. The university operates the app and has access to a corresponding portfolio of apartments for rent and lease to students. Via the app, the university grants each entitled party (potential or existing customers) access to the existing portfolio in exchange for the user's personal data and a fee. The tariff design is not currently the subject of discussion. The customer/student provides his/her personal data, also in order to have his/her creditworthiness checked. Via the app, the student gets access to the university's apartment portal and can view the available properties. It would also be conceivable to offer subletting options.

---

[11] Kletti/Schumacher (2014), p. 140.

If the student books the apartment, this includes services that the university offers. In addition to a cleaning service (e.g. final cleaning), this is an insurance package as well as a credit check. Some of these processes could be further divided and one could look at what other business opportunities are hidden behind them.

How much time is spent on this phase in the workshop depends on the total time available for the workshop. Even this investigation is worth it in order to understand one's own existing and possible products and services.

► In connection with a slowly emerging blockchain application, this sub-step looks at what you offer as a company and what "returns" you receive from your stakeholders. These returns are possible further elements of your blockchain solution.

4.6 Decision Path

After all this preparation, one feels that one has already found a valid approach to solving a problem.

But before you fall into the trap and at the end only find a blockchain solution for the sake of technology, it is recommended that you test yourself against a decision path.

Over the last 10 years, various models and approaches have emerged which are now presented below (according to their chronological order of development).

4.6.1 Decision Path According to Birch-Brown-Parulava

This model is based on the advantages of distributed registers (distributed ledger technology, DLT for short) and is recommended if the decision for a blockchain has already been made. This approach helps to find out which type of blockchain (public, private, consortium in connection with permissioned access or without) to use. In this model, the decision-making process focuses on the user groups and how data integrity is ensured (cf. Fig. 4.10).

The authors' motivation for this model is the idea of better organizing financial services on the financial market and establishing an institution-spanning ledger system (ledger). The communication between technology, companies and regulatory authorities in the world of financial services should be made easier by proposing a multi-layered architecture of a common main ledger.[12]

It must then be decided who can use this main ledger. As part of a top-down approach, the question arises as to whether everyone can have access to this main

---

[12] https://www.finyear.com/Blockchain-A-legacy-of-transparency_a36758.html; accessed on September 18, 2019.

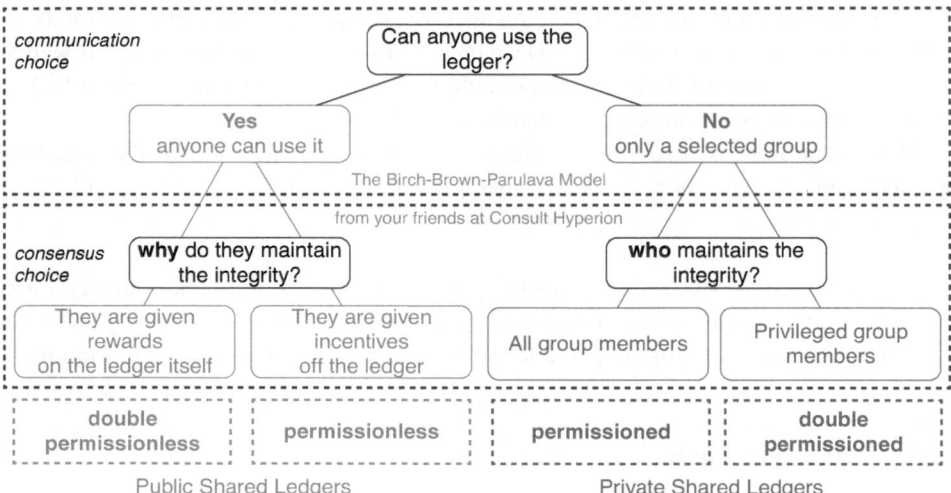

Fig. 4.10 Decision path according to Birch-Brown-Parulava. (with kind permission from © dwg-birch.com 2016. All Rights Reserved)

ledger and participate in the system, or whether it is only accessible to a few selected individuals/groups.

Answering this question then leads to the consensus level, on which the approaches with regard to public and non-licensed networks or private and thus licensed networks are already defined.

With the non-admission networks, the question arises as to why network participants should take care of the network and data integrity. The incentive mechanisms are divided into the following two categories: Either a network participant receives a reward for maintaining the main ledger or the participants receive incentives from the main ledger. In the former case, the decision leads to twice non-admission access to the network and thus to the ledger (main ledger) and is comparable to the mining of the Bitcoin block-chain.[13] The second case, in which the network participants receive incentives from the ledger, aims at a governance structure within this network.[14] The architects who choose this system can consider what incentives they offer in their own governance structures.

---

[13] The miners of the Bitcoin blockchain receive the agreed amount of Bitcoins for their maintenance of the blocks in the event of success.

[14] Standards define how to deal with each other, regardless of whether this refers to the digital or real world. In order for a governance process to function effectively, rules, responsible parties and participants must interact well with each other. For example, the rules should be aligned with the goals of the total participants, and those who can make rules should enforce positive and negative actions within this leadership structure.

This can range from co-determination rights with regard to necessary updates to validation mechanisms.

With the restricted networks, the question arises as to who within this network guarantees data integrity and thus also validates the transactions to be carried out. On the one hand, this can be all participants of this private blockchain, or on the other hand only a pre-selected group of people who are considered particularly trustworthy. The latter are then "Primus Inter Pari", the first among equals, because they have (again) special rights and duties. The initiators of the private blockchain select these validators and determine their rights and duties.

4.6.2   Decision Path According to Suichies

This approach is more about the question of the type of blockchain: Do I need a public or a private blockchain solution, a hybrid solution or maybe no blockchain solution at all?

You answer several questions along this decision path and "receive" one of the four possible outcomes: public blockchain, hybrid blockchain, private blockchain or no blockchain. The key questions here are: Are authors or those who add data known and trustworthy? If you know and trust the people who write data into your databases, then it may make sense not to use a blockchain approach (see Fig. 4.11).

You start by asking yourself whether you need a database at all. The fact that this question will now be answered with a 99% probability by all users of this model with "Yes" seems logical. However, if it should turn out that no database is needed, the decision path ends here, because it leads to the statement that no blockchain solution is necessary.

In the second step, you will be asked whether a shared write access is required. It will be asked whether different people are granted write access to the database (should). If this is not the case, the path ends with the recommendation not to use a blockchain solution. However, if you answer this question in the affirmative, you will be asked next whether the people with write access are known and trustworthy. This question is not so easy to answer. Yes, in most cases you will know the people to whom you have granted write access to your database. The question of trustworthiness, on the other hand, is assumed in the majority of companies—there is always a residual risk. Therefore, the inventor of the model questions this by asking whether the interests of the write-enabled can be seen as uniform. If you don't do everything yourself in your company, it is recommended to answer this question more critically and to go through the next step in the sequence.

If all inquiries so far have related to internal processes, this question is addressed to external third parties: Do you want/need to use a trusted third party? If you use such a provider and trust him, you can abort here and do without a blockchain solution. However, if you need a third party that you only trust to a limited extent (or not at all),

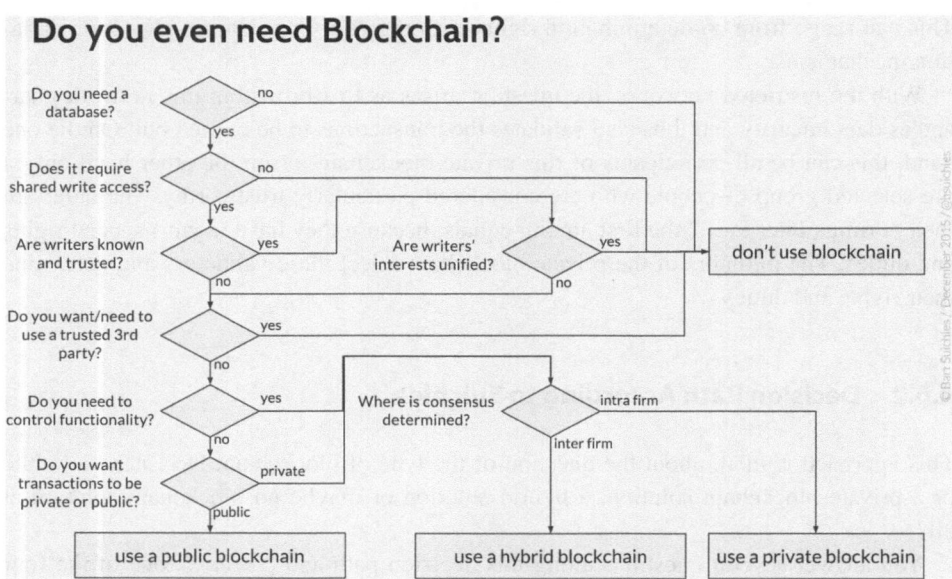

Fig. 4.11 Decision path according to Suichies. (From Suichies 2016: "When do you need a Blockchain?" With kind permission from © Suichies 2016 All Rights Reserved)

you will approach the question of whether and, if so, which functionalities[15] You want to control within the database. In the event that you waive (centralized) control mechanisms, the last question leads to the decision whether you want to make transactions within the database a) transparent and publicly visible to everyone or b) rather only allow selected users of the database. This allows you to derive whether you should pursue a public or rather a private blockchain solution. In the event that this decision path suggests a private blockchain solution to you, you must decide where the consensus is determined. If the consensus model is based on a purely internal company approach, then the private blockchain is the result of this decision path. If the consensus is to be created and applied between companies, then the recommendation is to choose a mixture of a private and a public blockchain. It is not apparent to what extent this hybrid approach mixes which elements from the private and public blockchain approach.

---

[15] In information technology, functionality (Latin functio means "execute") is the sum or any aspect of what a product, such as a software application or a computer device, can do for a user. The functionality of a product is used by marketers to identify product features and allows a user to have a range of functions. The functionality can be easy to use or not.

4.6.3   Decision Path According to IBM

In the IBM model, the needs of the market are clearly in the foreground. It defines the blockchain as a shared, unchangeable journal (cf. Fig. 4.12).

Interestingly, IBM "opens" the question of whether or not a blockchain solution is necessary with the question of the estimated transaction volume per second and high performance. If this claim is in the foreground, IBM recommends looking for alternatives. If you manage contractual relationships and your business logic[16] is more complex, IBM invites you to come and talk to IBM to jointly explore solutions.

Another indication of whether or not to consider a blockchain solution could emerge if identity plays a role within the contractual relationships and business logic you manage. Do you have to treat your transactions confidentially? Another indication to seek a conversation with IBM.

In the next iteration, IBM asks if the two previous aspects require market access. If not, then you can dismiss the idea of a blockchain. If you affirm the last question as well, IBM will lead you, with the hint that more than two parties are required, to the (ultimate) cost question: Do you want to reduce costs? This question is a no-brainer, just like the one that follows, in which you are supposed to decide whether you want to generate more visibility (in terms of Unique Selling Proposition) for your company. Both will be affirmed by companies, and thus IBM will be invited to a meeting.

Basically, a decision for a blockchain is based on the typical characteristics which are named in this decision path as follows:

1. No party can change, delete, or append a record without consensus, which makes the system valuable in order to ensure the unchangeability of contracts and other legal documents.
2. Smart contracts aim to provide a security that is superior to traditional contract law and to reduce other transaction costs associated with contract formation.
3. If everyone on an exchange can see the same ledger, it is easy to transfer an intention (or offer) by attachments. For example, in trade networks, all requests and bids for each network participant would be visible.
4. Blockchain networks allow every participant to create a tailored solution with its own proprietary business logic.

4.6.4   Decision Path According to Lewis

Another model is based on the questions of Antony Lewis (see Fig. 4.13).

---

[16]This term is used in software technology as an abstract term to separate the logic inherent in a task from the technical implementation.

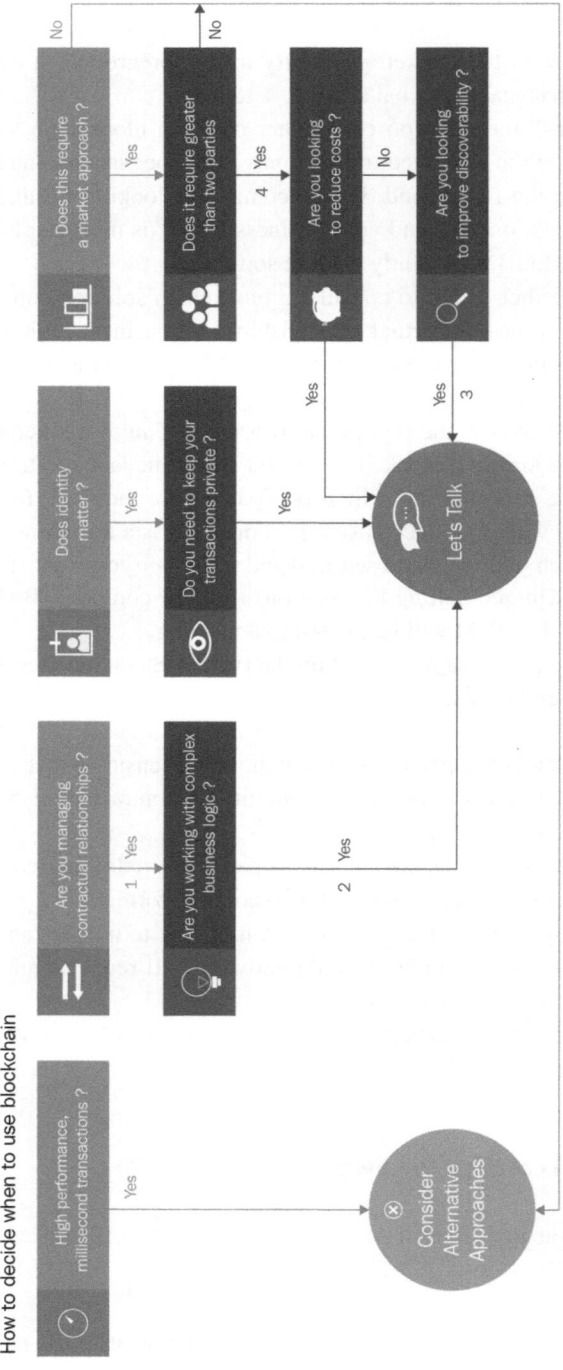

How to decide when to use blockchain

Fig. 4.12 Decision path according to IBM. (with kind permission from © IBM 2018 All Rights Reserved)

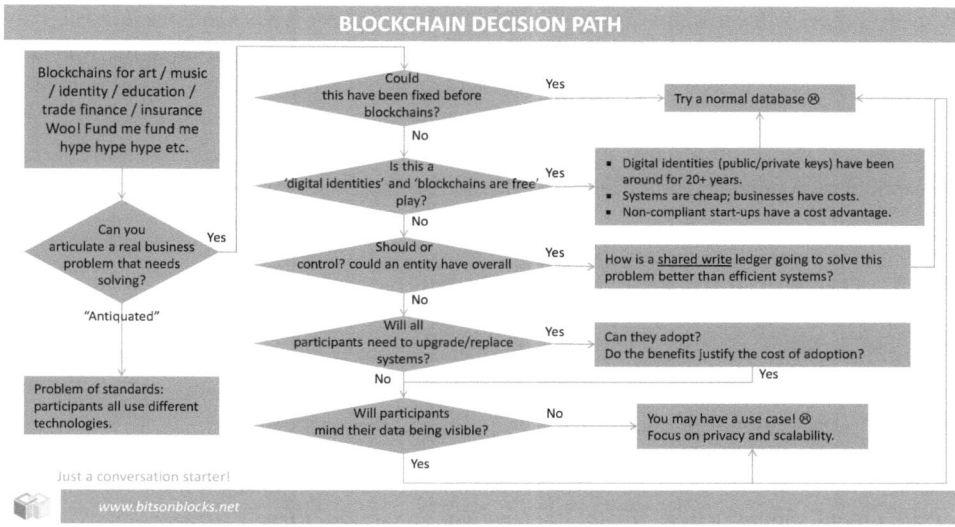

Fig. 4.13 Decision path according to Lewis. (with kind permission from Antony Lewis © www. bitsonblock.net 2018 All Rights Reserved)

The journey through these questions begins with a rather unusual and self-ironic statement that it actually has to be the ultimate idea.

But then Lewis brings you back to reality by asking you to name a real problem in the business world. He considers problems of different standards that are based on the fact that all participants use different technologies to be outdated.

Even if he packs this more as the beginning of a conversation about the topic, the question follows, if this can not be solved in other ways than by blockchain, that one of the conventional databases is recommended. Should this be the case, then one of the conventional databases is recommended.

With the subsequent question of whether it is the "digital identities" approach and/or "blockchains are free", this model points out that there have been digital identities with asymmetric key pairs (public and private key) for more than 20 years. Systems of this kind are inexpensive. On the other hand, companies incur more costs when they generate them themselves. Start-ups, on the other hand, could have a cost advantage. Traditional databases are better suited to process and process this rather well-known approach.

Unless it is "blockchain for free" or "digital identities", this model approach questions whether a company should or could have sole, overarching control? If this is answered with "yes", one is confronted with the question of how a shared write-ledger can better solve this business problem than existing efficient systems. Unless you have a good answer, the model then refers back to the use of traditional databases.

If the question of control in one hand/organization was denied, the model questions whether the participants have to update/replace systems? If so, can the participants take it over? And do the benefits justify the costs of taking over (whatever these costs may look like)? If it can be assumed that the participants have no objections to the transparency of their data, then it could be a worthwhile use case. In the further development of the business model, the focus should be on privacy and scalability.

However, if potential participants have objections to making their data transparent, this model recommends that you rethink the use of traditional databases.

4.6.5 Decision Path According to Meunier

The author of this model approach has built it as a check box. As already familiar, one sees oneself as a potential user of some questions to which it is necessary to find answers (cf. Fig. 4.14).

The following theses or assertions with respect to network, performance (power), business logic (in the sense of software technology) and consensus are made. The user of the model should decide whether he/she can agree with these theses or not. Meunier as the author of this model assumes that if one has less than seven agreeing statements, a blockchain solution is not worth it for the corresponding company because too many compromises and adjustments have to be made to make it "blockchain compatible" at all. But then it will be less of a real blockchain solution, but rather a shared traditional database.

To the assumptions:

Regarding the network:

- A significant number of participants will carry out transactions in the network (>100).
- It is neither necessary to trust the participants in the network, nor is there any reason to know the participants.

	Assertion	Answer	
Networks	A significant number of participants will be transacting on the network (>100)	Agree / Yes	☐
	You don't trust the participants in the network and you don't need / have to know them	Agree / Yes	☐
Performance	A limited amount of data needs to be stored for every transaction (a new fields)	Agree / Yes	☐
	The business process doesn't requires a high throughput (scalability)	Agree / Yes	☐
Business Logic	The business logic is simple	Agree / Yes	☐
	Privacy of transaction is not an important feature	Agree / Yes	☐
	The system will be standalone, it doesn't need to access external data or be integrated in IT legacy	Agree / Yes	☐
Consensus	No arbitrator shall be involved in case of a dispute	Agree / Yes	☐
	All participants can be involved in the validation of transaction (vs. only a group of known validator)	Agree / Yes	☐
	You need strictly immutability of the record (no amend & cancel, even by admin)	Agree / Yes	☐

Fig. 4.14 Decision path according to Meunier. (From Meunier 2016, "When do you need a Blockchain?"; With kind permission from © Meunier 2016 All Rights Reserved)

Regarding performance:
- A limited amount of data must be stored for each transaction.
- The business process does not require high utilization (scalability).

Regarding business logic:
- The business logic is simple.
- The confidentiality of transactions is not an important feature.
- The system will be standalone, and it will not need to access or integrate with external data or the existing IT environment.

Regarding consensus:
- In case of a dispute, no arbitrator can be involved.
- All participants can be involved in the validation of transactions (vs. only a group of known validators).
- Strict immutability of the dataset is a requirement (no change or termination, not even by the administrator, is possible).

Once again: The more you can agree with these assumptions, the more likely it is that this points to a blockchain solution.

4.6.6   Decision Path According to Wüst & Gervais[17]

These authors also deal with the question of whether a blockchain solution is actually necessary. As a starting point for their consideration, they distinguish between private and public or "permissionless" and "permissioned" blockchain variants. They point out that the use of a public or restricted blockchain is only recommended if several mutually mistrusting units interact and want to change the state of a system. In the context of these conditions, the participants of this system are not willing to agree on a trustworthy online third party and therefore need a blockchain solution (cf. Fig. 4.15).

This model is very clearly structured as a flowchart and starts with the question of whether you need to save any kind of status. If this is not the case, you can already leave the path here, because the question of a blockchain solution does not arise for you.

But if you still want to save a status (e.g. account balances, property rights, etc.), you have to ask yourself whether multiple authors can write to this status. These authors are defined as entities with write access to the blockchain/database and thus correspond to a consensus participant. Again, the question of whether your system should allow this or not is crucial. If the answer is "no", you will not use a blockchain type, but rather use traditional databases.

A "yes" leads you to the question of the use of so-called trusted third parties (trusted third parties; TTP). If you have planned this trusted institution in your business model/

---

[17]Wüst & Gervais (2017), p. 3.

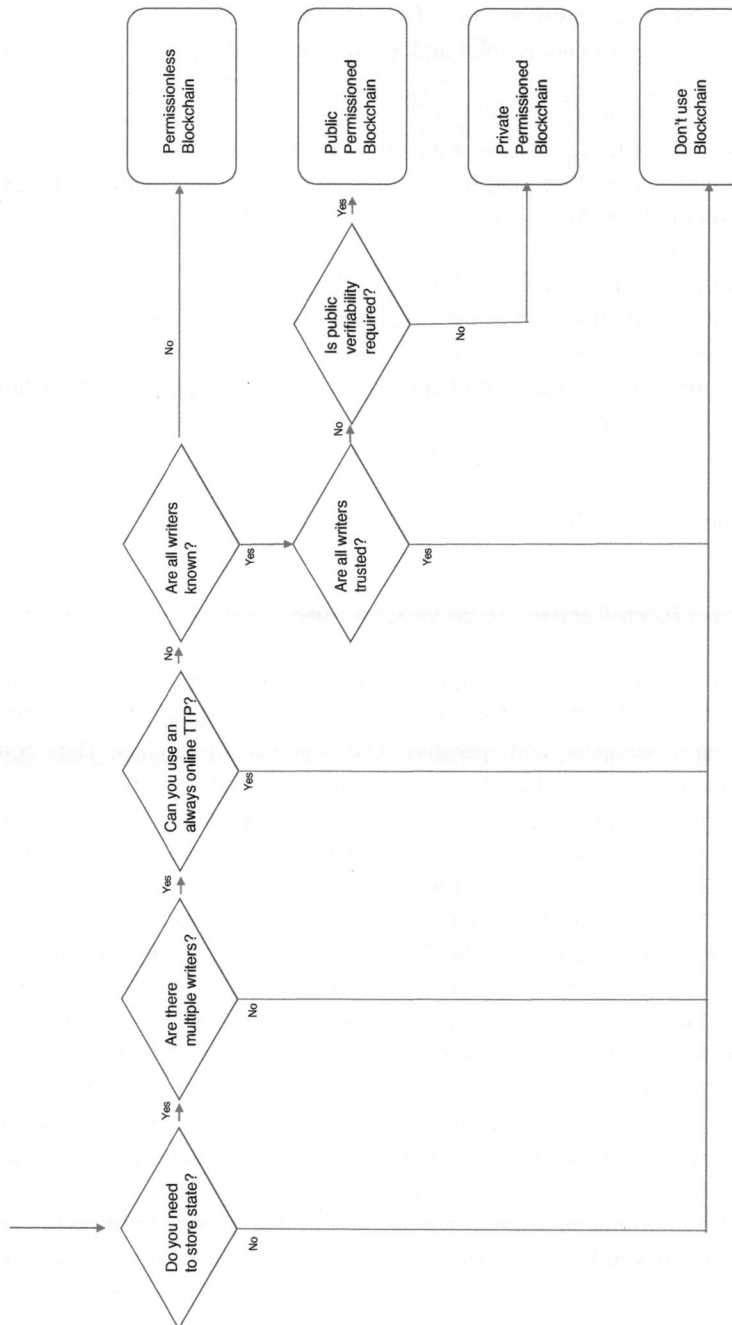

Fig. 4.15 Decision path according to Wüst & Gervais. (From Wüst and Gervais 2017; with kind permission from © Wüst & Gervais 2018. All Rights Reserved)

idea for a blockchain solution and the trusted institution is permanently online, then a blockchain solution is not recommended. Wüst and Gervais explain in their paper that a trusted third party can be available without being always online. In this case, the TTP can act as a so-called certification authority. From this assumption, the question is then derived of whether the authors/writers are known or unknown. If they are unknown, this leads to the recommendation to think about a public blockchain solution that allows everyone to read and check the validity of the stored data.

If the authors are known and trustworthy, a blockchain model is not necessary again. In the opposite case, that the authors are neither known nor trustworthy, the path leads via the question of whether a public verification is required, to the following two variants:

In the case of a "yes, a public verification is required", the model recommends a public blockchain variant with admission requirements for corresponding potential participants.

In the case of a "no, a public verification is not required", this procedure recommends a private blockchain that only allows a limited number of participants to read and use the chain.

4.6.7 Decision Path According to Peck[18]

Only a short time after the publication of the model by Wüst and Gervais, Morgan E. Peck leads through his decision path. Here too, it is only a question of clarifying on what basis an entrepreneur or other blockchain advocate comes to the assumption that he/she wants to use this type of data management. Peck refers to the fact that relational databases, which align information in updatable tables of columns and rows, are the technical basis for many services that we use today. The sole task of storing and updating entries is entrusted to one or a few units, which must be trusted not to manipulate the data. And, as Peck justifies his approach, this leads in consequence to the re-organization of data management by means of suitable technology (cf. Fig. 4.16).

The starting question of the decision path is—as already seen in the other approaches—a question that leads to the recommendation to consider a solution without blockchain technology if answered positively.

Since this flowchart is somewhat more confusing than, for example, the path of Wüst & Gervais, the following shows which questions one should ask oneself in order to come to a decision:

- After the initial question, the question arises as to whether more than one participant must be able to update the data.

---

[18] Peck (2017).

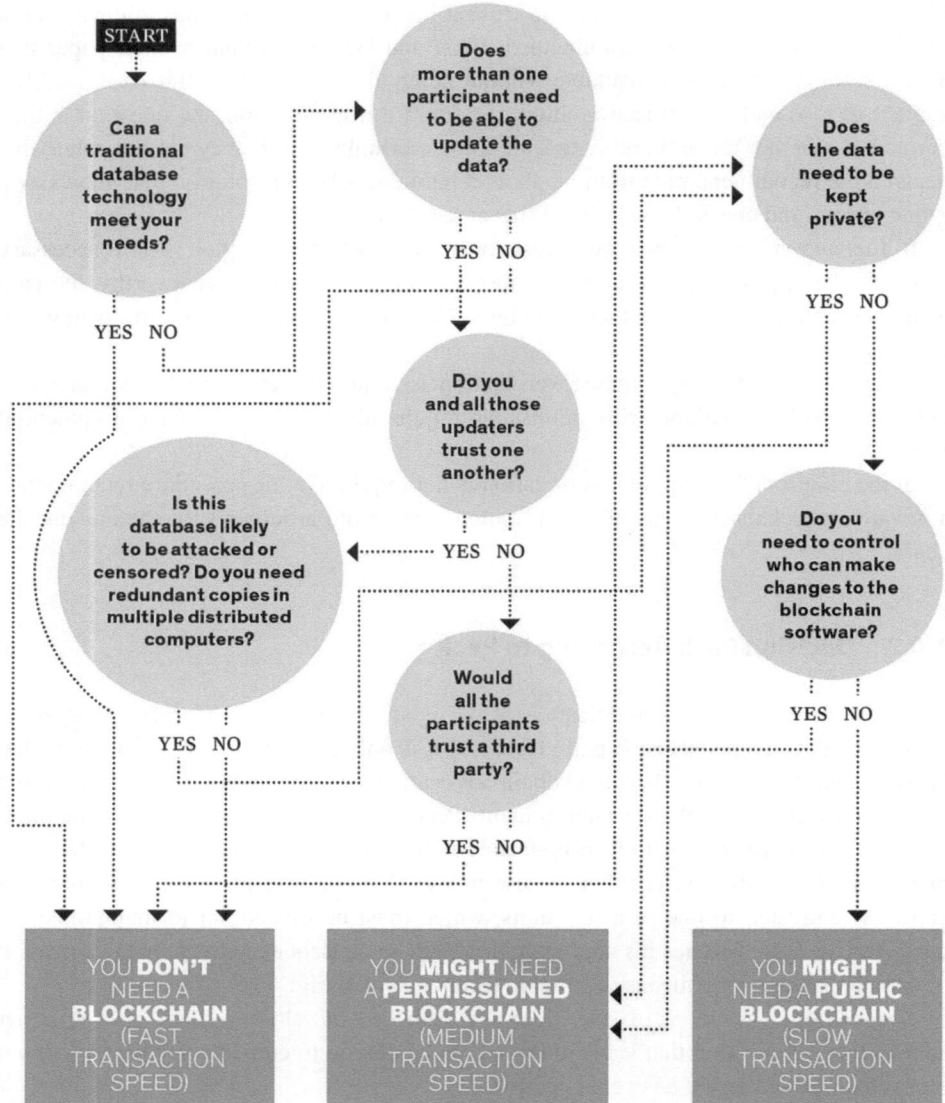

Fig. 4.16 Decision path according to Peck. (with kind permission from © Spectrum IEEE 2018. All Rights Reserved)

- Do these "updaters" trust each other?
- Is it likely that this database will be attacked or censored? Do you need redundant copies on multiple distributed computers?
- Would all participants trust a trustworthy third party?

- Are the data to be treated confidentially?
- Do you need to control who is allowed to make changes to the blockchain software?

Depending on which of these questions you answer with "yes" or "no", this will lead you to the decisions whether you need a blockchain or rather a high transaction speed. Or, you need a blockchain with participants who are given permission to participate. Within this approach, the transaction speed is lower than with traditional databases, but still higher than with the last variant of a public blockchain, which by system has a very slow transaction speed (note by the author: in 2017 this statement was true).

4.6.8   Decision Path According to United States Department of Homeland Security (DHS)[19]

The research team led by Dylan Yago uses a flowchart from the United States Department of Homeland Security (DHS) Science & Technology Directorate, which also examined the blockchain technology (cf. Fig. 4.17).

What is interesting about this approach is that the negative answers to statements and explanations meet, which alternatives there are.

Beginning with the question of whether a common, consistent [20] data storage is required, the right side shows the use of e-mail or spreadsheet as an alternative, provided that no historically consistent data storage is required.

Do you need more than one participant to store data in this data store? Yes? Then the consideration follows whether the data records stored in this data store are neither updated nor deleted. However, if only one participant stores the data, this is a strong indication against a blockchain application, because blockchains are usually used when the relevant data come from many different participants. Therefore, in the case of only one database user, no blockchain solution is necessary, but this approach recommends conventional databases. However, a warning is given: Please allow yourself a review of your use case.

Maybe there are records that were written once and never updated. Then you are back "in the game". Blockchains do not allow changes to historical data; this is well verifiable.

---

[19]Yaga et al. (2018), p. 42.

[20]In the sense of "Inconsistency itself": Consistency in database systems refers to the requirement that certain database transactions may only change data in a permissible manner. All data written to the database must be valid according to all defined rules, including constraints, cascades, triggers, and combinations thereof. This does not guarantee the correctness of the transaction in every respect that the application programmer would have desired (this lies in the responsibility of the code at the application level), but only that any programming errors cannot lead to a violation of defined database restrictions.

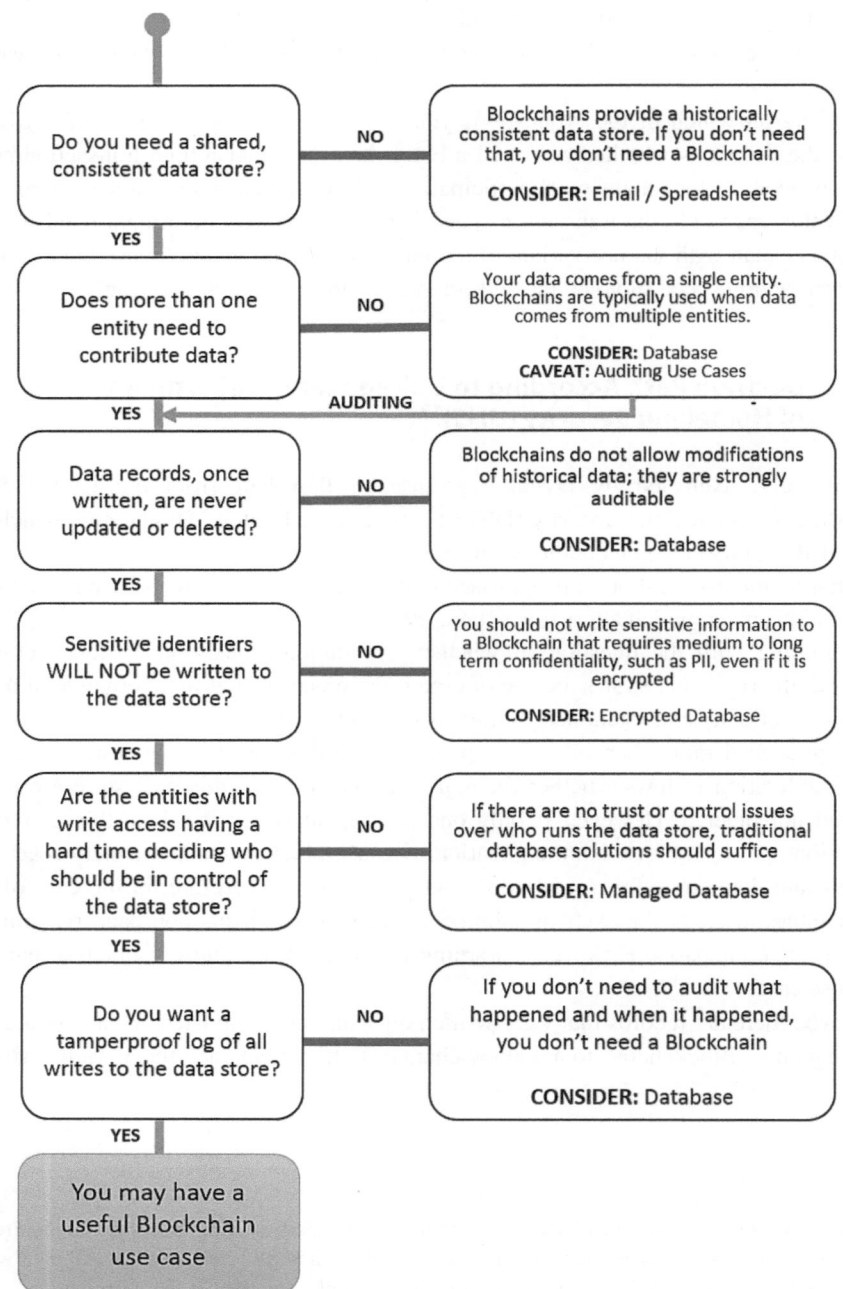

Fig. 4.17 Decision path according to DHS. (From "Blockchain-Technology Overview"; courtesy of © NISTIR 2018. All Rights Reserved)

If sensitive information from participants is not to be written and stored in the data store because it requires medium to long-term confidentiality, even if it is encrypted, then this model proposes encrypted conventional databases.[21] Otherwise you stay on the "blockchain path".

The model asks in the next step whether there are obstacles in the decisions for the write-enabled participants, who should have control over the data. If there are no trust or control problems about who runs the data store, a traditional database solution should suffice.

If you need a tamper-proof protocol of all write operations in the data store, you should consider a blockchain solution. If you don't have to check what happened when, you don't need a blockchain.

4.6.9   Decision Path According to Mulligan[22]

Author Cathy Mulligan is an expert and fellow of the World Economic Forum's Blockchain Council and published this approach in 2018 to make it easier for companies to make decisions. To do this, Cathy Mulligan analyzed how blockchain is used in a variety of projects around the world. In addition, she conducted interviews with selected CEOs with a research team and found that there are no more than 11 questions that companies need to answer to see if blockchain is a solution to some of their problems.

The route presented by Mulligan begins with the consideration of whether companies want to remove intermediaries from their business processes.

If a company also works with digital assets and can additionally create a permanent reference for these values, Mulligan recommends not to use a blockchain. In addition to this recommendation, Mulligan asks the question within her decision path at this point whether the respective company needs powerful, very fast (in the millisecond range) transactions. Since the 2018 existing blockchains (e.g. Bitcoin Blockchain, Ethereum Blockchain, IBM Hyper Ledger Blockchain) can not yet handle mass transactions, Mulligan advises against the use of a blockchain solution.

For the first time, the model shows here that this restriction still exists in 2018, but it is pointed out that within the community very intensively on the solution of this problem is worked. (Even the author of this book is of the opinion that it is only a matter of time until the transaction rate per second is as high on a blockchain as on a traditional database.) In an identical category falls the next question: Do you intend to store large

---

[21] Author's note: Whether this is a better alternative remains to be seen. The security requirements for a central data store must be extremely high in order to adequately protect sensitive data. It is much easier for hackers to break into a data store than the decentralized ledgers that are updated in synchronization.

[22] Mulligan (2018).

amounts of non-transaction related data as part of your solution? This is currently difficult to see with the existing blockchain solutions.

Unless a company does not focus on the previous questions, it is subsequently questioned whether the company wants to or has to rely on a trusted partner (e.g. with regard to compliance or liability reasons). This is the first indication that a blockchain-based solution is to be considered. The following question within the decision path almost sounds rhetorical in order to clarify whether the respective company, which is checking for itself the use of a blockchain solution, manages a contractual relationship or a exchange of goods. In connection with the questions already worked on, Mulligan suggests to do further research, because even if it does not seem so at first glance, there may still be potential.

Building on the existing knowledge, it is now queried within the decision-making process whether the company needs a shared write access to stored transactions. If a shared write access is required, it is obvious to ask whether the participants know and trust each other. Based on this, it is then to be clarified whether the interests and motives of the participants are well coordinated and uniform. Assuming that the corresponding company affirms this, Mulligan sees sufficient indications for a more detailed examination, since a blockchain application is probably value-creating.

With the questions at the end of her decision-making process, Mulligan directs the user of the process towards the type of blockchain: For Mulligan, an indicator for a public blockchain solution is if the network needs to be able to trace transactions publicly. At the same time, it is of secondary importance in this case that the network can control functions (e.g. updates).

The recommendation for a private and access-restricted solution results from Mulligan if the corresponding network values to control these functions (e.g. updates) centrally and to orchestrate them as well as to grant access to the network and thus to the transactions only to selected participants.

4.6.10 Gardner's Decision-making Process

Jeremy Gardner opens his question catalog by questioning whether you can name a real business problem that needs to be solved. This entry could also be seen in the model by Lewis—and in a modified form at IBM and Mulligan—which puts the focus on the needs of the market (cf. Fig. 4.18).

If you have identified a real problem, you are faced with the hypothetical question of whether this problem (if recognized) could have been solved before the blockchain era.

Then comes the usual question of the need for a database and the demand for whether there should be multiple write permissions. Are these write permissions known and trusted? Do the write permissions pursue uniform interests? Then a blockchain solution is rather to be rejected because it is not necessarily necessary to solve this business problem.

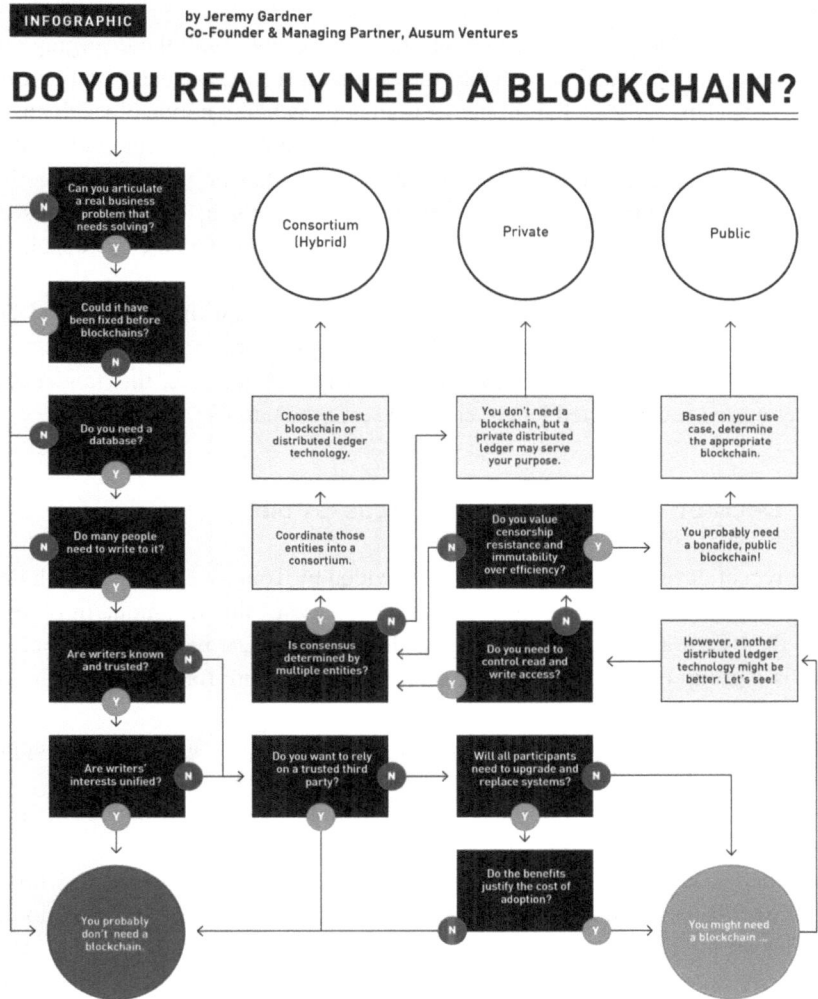

Fig. 4.18 Decision path according to Gardner. (with kind permission from © Gardner 2017. All Rights Reserved)

As in some other models, the consideration is raised as to whether a trustworthy third party might be used. If not, the question arises to what extent the participants in the network have to update the system. Since this is associated with costs, the question follows as to whether it pays off for the participants. It is therefore indirectly asked what rewards for the participants who update the system are conceivable.

Overall, this approach leads to the realization that a blockchain solution could be of interest.

In order to be able to assess this better, the author shows with the diverging questions which blockchain approach is probably the best to solve the business problem raised at the beginning. This can be a hybrid-consortium approach or a public blockchain solution.

Is it necessary within your business problem to control the write and read access? Who determines the consensus and its emergence? One or more/many participants? In this case, a consortium blockchain solution is emerging in which the consortium is organized according to the problem-solution approach using the best blockchain application available on the market.

Is censorship resistance and immutability more important than efficiency? This also means the use of a public blockchain in this model approach.

The use of a private blockchain is not expressly mentioned in this model approach, but rather a private distributed ledger can also serve the purpose.

4.6.11 Decision Path According to Koens & Poll[23]

Another recent development from 2018 is provided by Tommy Koens and Erik Poll, who compare a variety of models in their paper on the use of this technique in order to then provide their own approach. According to their approach, a blockchain is only needed if there is a group of unknown participants who are striving for consensus, as well as a need for a database.

Thus, these authors see their approach as a "schema for determining the appropriate database type" (cf. Fig. 4.19).

The entry into this model approach is very similar to that of Suichies (2016), Wüst and Gervais (2017) as well as the DHS Model (end of 2017).

Does the occasion and need exist that data is stored? Is there someone who has sole write access? Does the functionality need to be controlled[24] ? Is it possible to commission a trustworthy third party?

These questions and the resulting further steps have the mentioned models in common.

But now the extensions or the focus on "new" aspects and the resulting approaches begin:

Is the need present that transactions interact with each other? If this is not the case, the authors recommend a distributed database (and provide a specific provider right

---

[23] Koens & Poll (2018).

[24] The control of the database functionality can include the definition of rules for setting database permissions (e.g. create, save, delete), how the data is stored in the database (a relational database or an object-oriented database) or how the database can be queried (e.g. ServerSQL or MySQL).

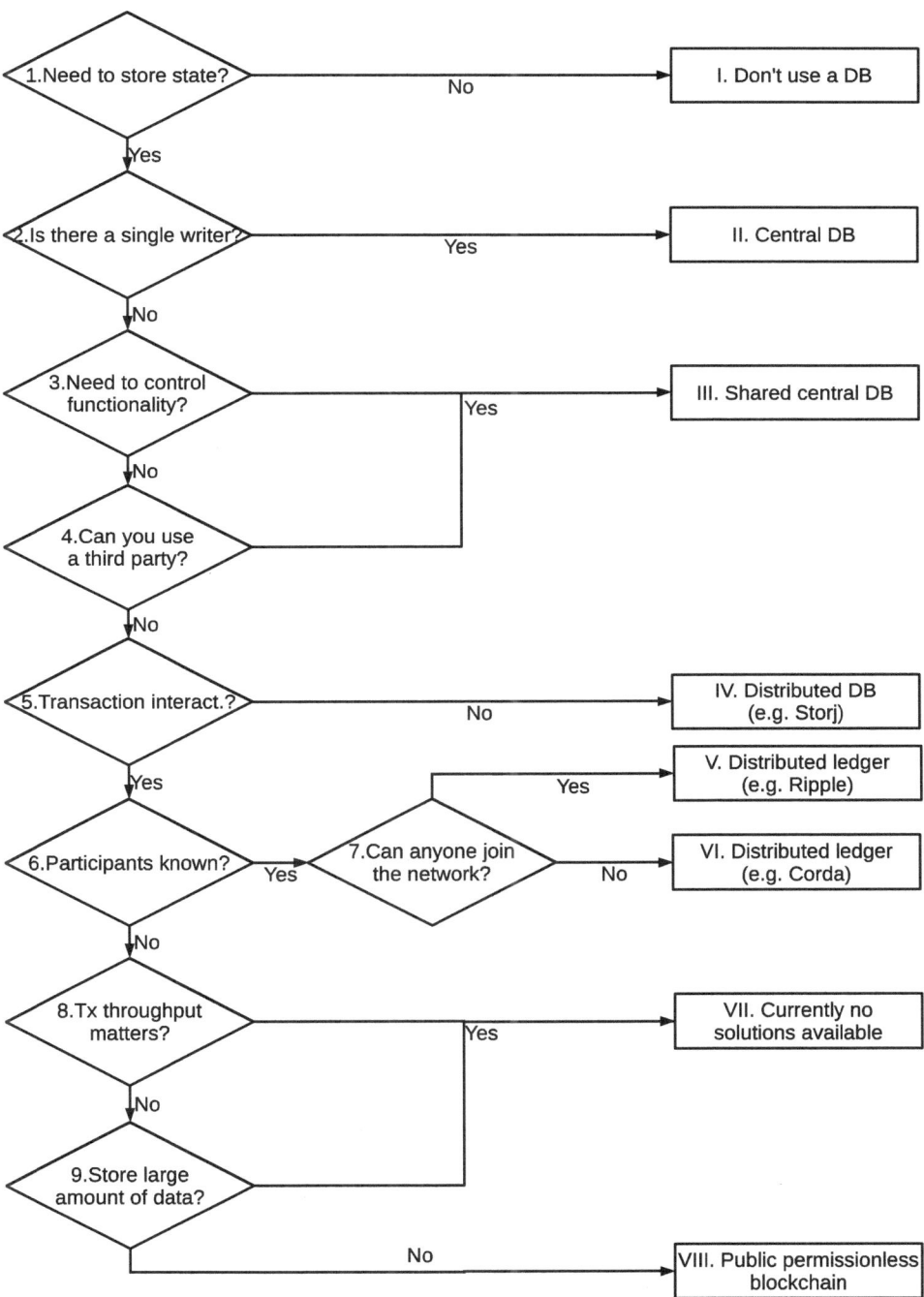

Fig. 4.19 Decision path according to Koens & Poll. (From Koens and Poll 2018, "What Blockchain Alternative do you need?"; with kind permission from © Koens & Poll 2018 All Rights Reserved)

away.) If, on the other hand, an interaction between the transactions is possible, this leads to the question of whether the participants who are to use this database within the network to be created are known. This question about the participants is refined by the question of whether everyone can join this network and thus become a participant. Since, as in all the models shown, the answer can only be given as "Yes" or "No", these authors derive two recommendations, depending on the degree of development: If the participants are known and can simply join the network, then a distributed register (in the sense of a common cash book) is recommended, for example the payment network Ripple. If, on the other hand, the participants are known but there are restrictions in the form of whether and to what extent one can join the network, the authors recommend another approach to a distributed network that can take these properties into account, for example Corda.

Finally, the questions are raised as to whether both the transaction throughput (that is, how many transactions can actually be processed per second) and the storage of large amounts of data are of high importance, accompanied by the observation that none of the existing blockchain solutions currently meets these requirements. However, significant improvements in scalability are to be expected in the near future.

4.6.12 Overview of the Decision Models

With the use of a decision path, you can clarify whether and, if so, what type of blockchain you should use for your application.

The so-called Decision Path Models are captured here for the purpose of comparing the differences and similarities.

Werner et al. add frequencies in terms of number/occurrence of the result to this tabular approach.[25] Table 4.2 does not use frequencies, but rather examines the models using the following criteria:

Focus: what does this model approach pay special attention to
Center/Center of Gravity: for which user group/industry is this model approach suitable
Expression: what recommendation does the model make based on the chosen approaches

4.6.13 Summary Interim Conclusion

The blockchain technology is still relatively new and many companies are looking for ways to integrate it into their business. The fear of missing out on this technology is

---

[25] Werner et al. in Fill/Meier (2020), p. 25.

Table 4.2 Comparison of decision paths

Path according to	To be found in	Focus	Center of gravity	Expression
Birch-Brown-Parulava	4.6.1	User groups and data integrity	Financial market	Choice between public and private blockchain; Incentives for participation in the system
Suichies	4.6.2	Blockchain Type	Access Rights	Choice between public, hybrid and private blockchain
IBM	4.6.3	Market Orientation	Cost Orientation.	Promotion of IBM solutions
Lewis	4.6.4	Real existing problem	Creating uniform standards	Control mechanisms; Cost aspects
Meunier	4.6.5	Question your own mind-set	Process-orientation	Check-box to determine company-specific awareness
Wüst & Gervais	4.6.6	Suspicious Interactors	Fundamental Questions on the Topic of "Trust"	Selection between public, hybrid and private blockchain
Peck	4.6.7	Reorganization of data management	Avoidance of data manipulation	Transaction speed, choice between public, hybrid and private blockchain
DHS	4.6.8	Alternatives	Data Storage	Checking other database solutions
Mulligan	4.6.9	Eliminating Intermediaries	Shared Writing	Transaction speed; amount of data to be stored
Gardner	4.6.10	Real existing problem	System access	Selection between public and hybrid blockchain
Koens & Poll	4.6.11	Database type	Transactions between databases	Transaction speed; permeability of different databases

quite large and many organizations approach the problem of using this technology "somehow". This leads to disappointment and frustration because blockchain technology is not universally applicable.

With the idea of the so-called decision paths, which are to be understood as a guideline for decision-makers, it is possible to approach the topic. These questions must be answered. It can be seen from the variety that the different approaches to the overarching question of whether the use of a blockchain is actually meaningful and "value-promising" differ. It can be seen that of the eleven models presented, five have chosen data storage as a starting point (cf. Suichies, Wüst & Gervais, Peck, DHS and Koens & Poll). Four approaches focus on the market and its needs in order to test their own idea for market acceptance (Lewis, IBM, Mulligan and Gardner). The model by Birch-Brown-Parulava focuses initially on the question of access in order to then show the various possibilities. Meunier formulates requirements to justify the implementation of a blockchain.

All the different models presented have their advantages and disadvantages. However, most of these models are reduced to presenting a decision algorithm that, through questions about the analyzed use case, finally leads to the question of whether or not a blockchain should be used. The binary approach is visible in all these models: Yes or no. However, all models lack the detailed analysis of the decision-making process in order to filter out (possibly also partial) the process that brings the decisive added value to the company.

Therefore, it is ultimately about the following core questions:

- Who has access to your data and how?
- Which data in your systems can be changed or even manipulated by whom and with what consequences in your company?
- Which processes involve too many stakeholders?
- For which processes/services do you pay too much?
- Which intermediary bothers you the most? Why?
- Does real-time monitoring of activities between, for example, regulatory authorities and your company need to be possible?

With this knowledge "in your luggage", you are able to carry out a critical self-analysis.

▶ Please question yourself and also within the team the motivation why you believe in a blockchain solution. Which type of blockchain with which orientation helps you best to answer the core questions posed? Have the courage to also reject a blockchain solution if no added value is recognizable. Maybe you are not that digitalized with some processes yet. The implementation of a blockchain approach could mean that you go from step three to step one!

After you have gone through a decision-making path and answered the core questions, you can transfer the processes and properties that are increasingly crystallizing into the value-added analysis.

▶ You have looked at the different approaches to decision paths and have certainly already found a model that is suitable for you and your business idea while reading. Please write down the questions again for the approach you prefer and discuss them as a team. Compare your "story mapping" card against it and check your findings and previous results.

4.7   **Value Analysis**

Your story mapping card has given you first-hand knowledge of which of the values you have classified as relevant (or interesting) can also represent digital values. But how can you decide which of the values and value promises, as well as processes and activities, is the most promising approach? The value analysis, or point rating method, helps to select alternatives by breaking down the overall problem into sub-aspects.

Kühnapfel points out that we tend to simplify complex, multi-layered issues. The consequence of this is, on the one hand, a higher (than necessary) error rate and, on the other hand, the tendency to stay in order not to have to make a decision.[26] For this reason, you should always go through the following steps of this analysis!

A value analysis is always useful when

- many aspects have to be considered,
- assumptions have to be made about qualitative and quantitative aspects,
- it is difficult to determine an adequate sequence,
- the risk of wrong decisions should be reduced and
- decisions should be objectively and comprehensibly prepared (e.g. for the management and/or the supervisory board).

In principle, a utility analysis is carried out as follows:[27]

1. Selection of the decision problem (in our case, since you have created the product and service map).
2. Selection of the alternatives (also found through the product and service map).
3. Concretization of decision criteria.
4. Subdivision of criteria and pre-selection.
5. Weighting of criteria.
6. Evaluation of alternatives.
7. Determination of utility values.
8. Decision.

By the work already done, you have knowledge about points 1 + 2 available. Now it is time to exactly define the decision criteria. For this purpose, a strictly hierarchical goal system is suitable, in which higher-level goals, which are to be kept more general, are formulated in order to derive sub-goals and objectives from them. Sub-goals are always part of the next higher goal (cf. Fig. 4.20)

---

[26] Kühnapfel (2019), p. 2.

[27] Kühnapfel (2019): p. 6.

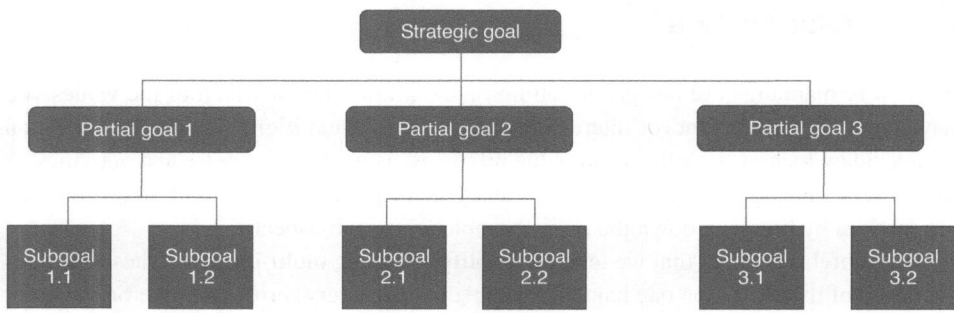

Fig. 4.20 Target pyramid

Using the continuing example of the university, the target pyramid could be set up as follows, for example:[28]

Strategic overall goal: More efficient administrative processes with
Subgoal (1): Reduce process cycle times,
Subgoal (2): Personnel relief through automation,
Subgoal (3): Cost reduction.

These subgoals can now be used to determine sub-goals. For this purpose, only sub-goal (1) is now divided into two sub-goals, to serve as an example. Thus, sub-goal 1.1 could be "Digital payroll (no paper prints)" and sub-goal 2.2 "digital certificates".

After you have discussed and defined your overall objectives and sub-objectives in the group, you now need to set up target criteria and evaluation criteria. There are so-called must-criteria (i.e. indispensable or also known as KO criterion)[29] as minimum or maximum conditions as well as desired criteria that should be met. It is recommended to also precisely define for desired criteria when the criterion is completely, partially, or unacceptable met.

As the fifth step in this process, you weight the found criteria. Not all defined criteria are equally important in order to achieve the overall benefit. The determination of the weighting will always be of a subjective character and should therefore be coordinated within the group (cf. Fig. 4.21).

The weighting is indicated along the importance of the individual criteria and can be set from 1 (not important) to 5 (very important). It is also possible to have a percentage weighting. However, in this case the sum of all individual factors must add up to 100%.

---

[28] Author's note: These are fictional, non-university-coordinated assumptions; they serve only for illustration.

[29] Note: any alternative that does not meet a must-criterion is eliminated.

Processes	Name P1					
	Name P2					
	Name P3					
	Name P4					
	Name P5					
	Name P6					
	Name P7					

		Rating scale				
Criterion	Determination method			2	1	
Criterion K1						
Criterion K2						
Criterion K3						
Criterion K4						
Criterion K5						

Fig. 4.21 Nutzwertanalyse I

In order to weight overall objectives and sub-objectives equally, the total weight is distributed step by step to the corresponding target levels.

Next, the evaluation of the alternatives takes place (step 6). For this purpose, points are assigned to the alternatives, which indicate how well or badly they meet the target criterion. Evaluation systems such as school grades from 1 (very good) to 6 (insufficient) or a ranking from place 1 to place n or point values 0 to 3 etc. can be used (cf. Fig. 4.22).

▶   **Tip** The award of points is not always easy and will trigger intense discussions within the team. In practice, the problem of possibly working with too much bias is solved by dividing the evaluation into three categories using scenario technique: "best case", "base case" and "worse case" as assumption under best conditions, assumption and normal conditions as well as assumption under bad conditions.

In this way, the range in which the utility values lie can be delimited. You can find the overlap levels and would therefore have relatively safe overall assumptions.

The cross table can now be used to determine the values and filter out the process in which the highest value is achieved for the company (cf. Fig. 4.23).

You will find the value table shown in Figs. 4.21, 4.22 and 4.23 as an Excel sheet in the online version of this chapter as supplementary material for download.

As the example of the university is continued, you can see Fig. 4.24.

Criteria weighting by pairwise comparison					
Criteria	Criterion K1	Criterion K2	Criterion K3	Criterion K4	Criterion K8
Criterion K1		n.a.	n.a.	n.a.	n.a.
Criterion K2			n.a.	n.a.	n.a.
Criterion K3				n.a.	n.a.
Criterion K4					n.a.
Criterion K5					
Sum	0	0	0	0	0
Meaning	0%	0%	0%	0%	0%

Legend:
2 - Column more important than row
1 - Column and row equally important
0 - Column less important than row

Determine evaluation criteria							
Criterion/Process	Name P1	Name P2	Name P3	Name P4	Name P5	Name P6	Name P7
Criterion K1							
Criterion K2							
Criterion K3							
Criterion K4							
Criterion K5							

Fig. 4.22 Utility analysis II

Determination of value in use								
Designation	Name P1	Name P2	Name P3	Name P4	Name P5	Name P6	Name P7	Meaning
Criterion K1	EMPTY	EMPTY	EMPTY	EMPTY	EMPTY	EMPTY	EMPTY	0%
Criterion K2	EMPTY	EMPTY	EMPTY	EMPTY	EMPTY	EMPTY	EMPTY	0%
Criterion K3	EMPTY	EMPTY	EMPTY	EMPTY	EMPTY	EMPTY	EMPTY	0%
Criterion K4	EMPTY	EMPTY	EMPTY	EMPTY	EMPTY	EMPTY	EMPTY	0%
Criterion K5	EMPTY	EMPTY	EMPTY	EMPTY	EMPTY	EMPTY	EMPTY	0%
Overall result	0	0	0	0	0	0	0	
Weighted result	0	0	0	0	0	0	0	
Ranking	1	1	1	1	1	1	1	

Note on error message:
EMPTY = "Determine rating criteria" and "Rating scale" are incomplete
ERROR = Input "Determine rating criteria" and "Rating scale" do not match

Fig. 4.23 Value analysis III

You can see from the table that, according to the ranking, the provision of the final certificate is the worst process and, from the company's or university's point of view, the one that creates the most value when it is improved.

► The use of blockchain technology is supposed to create value for the company. The value-added analysis makes it possible to filter out those processes that have been identified from the previously selected application cases and for which blockchain technology can create real value. In addition, the assessment serves later to determine the value creation.

4.8　The Morphological Box

The morphological box is a (further) creativity technique to obtain ideas for questions and solutions in a structured manner. Even if this is mostly a two-dimensional model and only rarely a three-dimensional box, this technique helps to approach unusual solutions.

Morphologically means "regarding morphology, based on it, belonging to it; regarding the external shape, form, structure … ".[30] The inventor of this technique, Fritz

---

[30] Duden (1990), p. 515.

Processes	Provide proof of graduation
	Confirm student status
	Document research results

Criterion	Method of investigation	Scoring scale			
		3	2	1	0
Transparency	Survey	All	External + internal	Internal	None
Falsification security	Test	Very high	High	Low	Very low
Cost	Estimate	< 1€	1 - 10 €	10 - 100 €	> 100 €

Criteria weighting by pairwise comparison

Criteria	Transparency	Anti-counterfeiting	Cost
Transparency		1	0
Anti-counterfeiting	1		1
Cost	2	1	
Total	3	2	1
Importance	50%	33%	17%

Determine rating criteria

Criterion / Processes	Provide proof ...	Confirm student...	Document research results
Transparency	Internal	External + internal	External + internal
Anti-counterfeiting	Very low	Very low	High
Cost	10 - 100 €	< 1€	< 1€

Determination of value in use

Designation	Provide proof ...	Confirm student...	Document rese...	Importance
Transparency	1	2	2	50%
Anti-counterfeiting	0	0	2	33%
Cost	1	3	3	17%
Overall result	2	5	7	
Weighted result	0.67	1.50	2.17	
Cost	3	2	1	

Fig. 4.24 Value analysis using the example of the university

Zwicky, recommends the use of the morphological box when, for example, it comes to product improvements, product updates or product relaunches.[31] With the help of the morphological box, a problem is broken down into individual parts in order to combine and assemble these fragments in a variety of ways until a plausible solution has arisen for the team.

Even if some sources hold that this method helps to show the totality of all conceivable solutions to a question, this claim does not have to be fulfilled. Only the decomposition of the given question into individual parts, into so-called variable design parameters, forces the users of this technique to define these characteristics. Sounds simple, but it is not.

---

[31] Mittelmann; https://www.artm-friends.at/am/km/WM-Methoden/WM-Methoden-285.htm, accessed on 01.05.2021.

Table 4.3 Structure of the morphological box

Parameter/ Feature (What?)	Characteristic I (How?)	Characteristic II (How?)	Characteristic III (How?)	Characteristic IV (How?)

In addition, execution options are listed in terms of variations. Arranged as a table, a matrix is created (Table 4.3).

Solutions are created by choosing an expression from each characteristic cell and connecting them with each other by lines.[32]

It can now be very helpful to take the freedom to provide the process, the product or the service that you have filtered out with various characteristics and expressions.

According to the example of university degrees that has accompanied us so far, you can trace how this creativity technique is used.

From the work steps you have carried out so far within the mentioned workshop tools, you have a good description of the existing situation and the resulting question.

According to the example of the university degrees to be secured by means of a blockchain, we have found that the aim is to create a university degree that is tamper-proof and counterfeit-proof. Every employer can check by means of the hash value whether the certificate submitted by the applicant is the original.

In the next step, you determine the parameters of the "problem" in independent features. You break down your product idea by approaching the "what" (parameters, components, properties, etc.) with questions about the problem. You are welcome to shoot beyond the target here and also to record initially completely absurd features. The "magic" of the morphological box will develop through the interaction of features and characteristics. For example, with a logo, the parameters could be as follows: image elements, word elements, pattern elements, core product elements.

Now transfer your characteristics/parameters to the table.

Then an expression is sought for each parameter and entered accordingly in the table.

And now the "fun part" begins: by combining the individual characteristics (the "how") with the characteristics (the "what"), you get new solutions. You illustrate this combination by connecting them with lines and making them visible.

Finally, you should discuss your solution as a team.

The accompanying example of blockchain-based certificates in this book is shown in Fig. 4.25 as a morphological box.

---

[32] Pricken (2007), p. 228.

Feature/ Parameter	Characteristic I	Characteristic II	Characteristic III
Material	Pure handmade paper	Cheque card chip	Plastic
Logo	Typical HTW logo	Symbols, matching the courses of study	Without logo
Shape	DIN A4	Micro card (12 x 15 mm)	Credit card format
Binding potential (to the university)	Moderate	High	Unimportant
Technology layer	Watermark	Data storage on chip	Blockchain technology
Target group	Parents & Family	Students & employers	Nerds
Additional function	Representative purpose	Cloud and internet access	None

Fig. 4.25 Morphological Matrix at the Example "Storage of University Certificates" I

You will find here parameters/characteristics as well as expressions. In Fig. 4.26 the most important expressions are connected to the characteristics. The result looks (exemplarily) as shown in the table.

Please note that this is a proposal for a solution.

In order to achieve satisfactory results, in addition to a little practice, a sufficiently deep knowledge of the process/product to be revised or created is also required.

It is also helpful if you determine the parameters logically independent of each other when defining them. This allows you to have a greater variety of combinations. Pay attention to the determination of the parameters that they do not describe subordinate

Feature/ Parameter	Characteristic I	Characteristic II	Characteristic III
Material	Pure handmade paper	Cheque card chip	Plastic
Logo	Typical HTW logo	Symbols, matching the courses of study	Without logo
Shape	DIN A4	Micro card (12 x 15 mm)	Credit card format
Binding potential (to the university)	Moderate	High	Unimportant
Technology layer	Watermark	Data storage on chip	Blockchain technology
Target group	Parents & Family	Students & employers	Nerds
Additional function	Representative purpose	Cloud and internet access	None

Fig. 4.26 Morphological Matrix at the Example "Storage of University Certificates" II

details and that the parameters can apply to all conceivable solutions. And, give yourself and your team enough time to create this matrix. It can sometimes take several days. You can derive valuable insights from it. The matrix supports the innovative potential discovered so far by you and your team with additions and expressions.

4.9   Business Model Canvas[33]

You have found one or possibly even two processes/use cases from the value analysis that urgently need a solution in the setting you have chosen. You can now map your result even further with the Business Model Canvas to better align the overall context.

Think systematically about your business model!

Alexander Osterwalder's Business Model Canvas has nine fields to describe the business idea compactly on a "canvas". As with a painter's canvas, you can "overpaint" this canvas again and again. With Post-its, keywords can be inserted into the fields and also quickly removed if it is found during the team discussion that the term does not fit or does not fit here (cf. Fig. 4.27).

Before you start filling this "canvas" with terms, please ask the following questions as a team:

- *How do you win customers?*
- *After you have won a new customer, how do you plan to identify yourself with this customer and manage the relationship (if at all)?*
- *How do you burden your customers? What is your revenue model?*
- *How much do you demand from your customers? Can you calculate your sales for the next month, the next quarter and year?*
- *What assets are available to you or under your control?*
- *Who are your most important partners?*
- *What key activities do you need to perform to fulfill your promise of performance?*
- *What are your fixed costs?*
- *What are your variable costs? Can you calculate your total costs for the next month, the next quarter and year?*
- *Does your sales forecast show increased profitability towards the end of the forecast period?*

So you can get started and turn to the model.

You "read" the map of the model from right to left, so the first field to consider is the Customer Segments field. Here you should ask yourself which customers you serve in your company?

---

[33] Osterwalder (2004).

Design your own business model

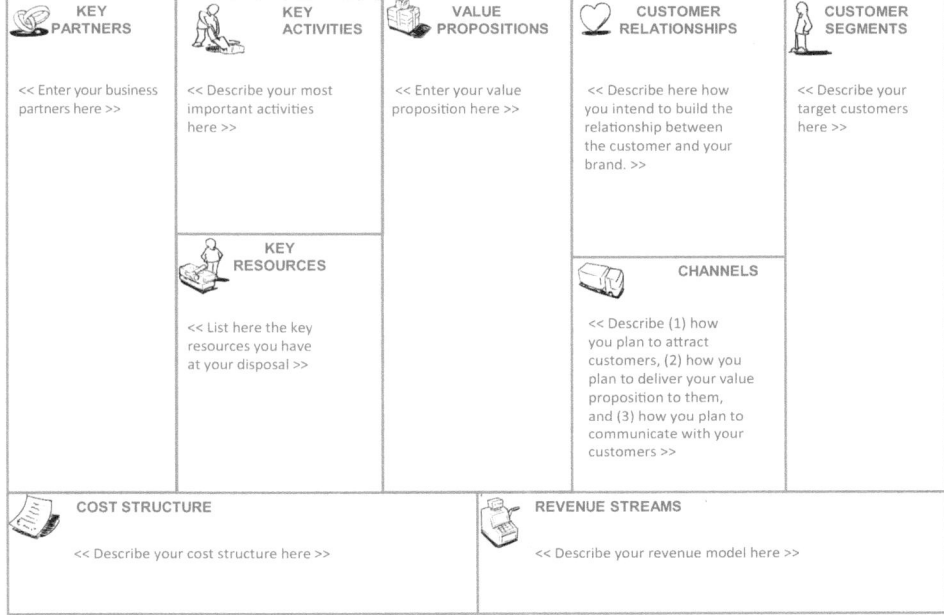

Fig. 4.27   Business-Modell Canvas according to Osterwalder

Let's look at the example of the university again, we can say that there are three customer segments that can be named as target groups:

a) the students,
b) the economy,
c) and, at a higher level, society as a whole.

If the teaching and the course offer are attractive, many students will choose the university and a corresponding degree. There are, inter alia, rankings for this, which can also be seen as an incentive for the teaching staff.

The economy expects well-trained graduates who meet the requirements.

On a higher, meta-level, but nevertheless not to be neglected, is society. The higher the educational level of a society, the better the economy of a country.[34]

We therefore have three different customer segments and three different tasks here. For the different segments and tasks you have a different performance promise each.

---

[34] OECD (2019), Education at a Glance, p. 77.

After you have found out for yourself who your customers are and what you want to offer them (that is, which service or which product), then you have to think about how you want to reach your customers or how do your customers want to be reached through which channel? Which channel do my customers want to communicate with me as a company or organization? How should they receive their goods? Is this a real or digital channel? Are we talking about stores in the sense of stationary sales units or are we talking about direct or indirect distribution channels? Please ask yourself how the customer segments identified by you want to be reached.

To supplement the example of the university, it can be said that it is no longer just about the physical presence of the university. Online courses and e-learning, etc. are becoming increasingly important to reach the segment of students. Universities tweet their messages and events just like profit-oriented companies. University marketing is now a subject of study. The task is to transfer the findings of consumer and industrial goods marketing to universities.

The next field deals with the customer relationship to be established. Perhaps your business model is very strongly based on personal relationships. This would then have further consequences for the other areas in this model. If you build a very person-related business model, the focus is on this personal level, but such an approach is not very scalable. Take, for example, one of the large Internet platforms such as Amazon, Alibaba, Google, etc. Do you think these companies have a personal relationship with us as customers? Maybe not personally in the sense of a company that knows every customer personally. But these companies have a personalized relationship with us because, for example, they understand our buying behavior and can recommend books or electronic devices, etc. to us. This type of company builds an automated relationship with its customers that is based on systems, computers, and servers. A completely different approach is pursued here, which is much more scalable.

You can also use the Customer Relationship field to describe how you win and retain customers. Customers must first be won, but once you have won them, it is important to find out how you will keep them. And if you don't have enough customers, you're "damned" to grow—or disappear from the market.

Therefore, the next important question is what people or companies you serve are really willing to pay for? What do you think your customers want to pay for something they buy? Do they want to pay for the subscription or rather a license fee. What are the necessary price mechanisms? Take Google as another example: Google auctions its search terms for advertising. They don't just sell them. This makes them much more successful than if Google would only sell them.

In the case of the university example, it is not so easy to explain how the customers are willing to pay. There are state universities whose offer is free of charge for the students despite the highest level. Only a administrative fee has to be paid. For the further costs the state and thus the society pays. But to assume that universities are accountable is wrong. But this is not the right place to explain the fiscal policy of the federal government, the states and the municipalities.

You can see that so far the right side of the model has been illuminated more closely. So the part that shows what value you create for whom and how you deliver it. And in addition, how you capture the value for yourself as an organization.

Now let's turn to the left side. Here it is questioned how the value is created. What is actually needed to create the value that the right side of the Canvas promises? Questions arise about the resources such as, for example, whether factories are needed, or maybe a brand/brand name? Does it need intellectual property or a server? What are the most important things that are needed to create this value?

The self-critical question of what needs to be done to really be good in our own business model follows. Which activities do we need to pay particular attention to? Is it marketing in sales, research and development, or perhaps the administration of servers? The driver behind these questions is to find out which activities are crucial for our business model. And: Can activities be outsourced? For this, partnerships can be entered into. This raises, in addition to the offer (from the right side), the question of who the most important partners are that can use your business model. This can turn your business model into something more powerful.

Zynga, a social gaming company, is an interesting example of this. A large part of its business model is built on the back of Facebook. This partnership has helped Zynga to scale so quickly and be used by more than 100 million players.[35]

If you know all three elements on the left side of your canvas, you can quickly find out what your costs are. Overall, you get an overview of how to create, deliver, and capture value in your business, and you can explain the logic of your business model.

Test the model in your team and fill in the fields according to your business idea.

Then rethink your business model:

Does it work? Do the revenues exceed the costs? Try to estimate revenues and costs for the next month ("or the first month of sales" if you are initially without revenue).

What are the risks to your business model? Which parts of your business model are most important to your company in order to grow profitably?

Are there things that can or should be changed to strengthen the business model or reduce the risk?

It is then recommended to check each answer in the canvas model using the following criteria in the table below (see Fig. 4.28):

- How do you know that your assumptions are correct or true?
- Have you made an assumption or do you have solid evidence, for example in the form of documented facts?
- In the cases where you have facts, mark the answer as "fact" and make a note of your evidence.
- In the cases where you have made assumptions, mark the answer with "assumption".

---

[35] Golem Plattform (2019) See https://www.golem.de/specials/zynga/ accessed on September 12, 2019.

Clarify the assumptions for your business model

Use the questions asked and check which ones can be answered with assumptions
and which of the questions can be answered with facts.

No.	Question	Fact/Facts	Assumption
1			
2			
3			
4			
5			
6			

Fig. 4.28 Check questions for the Business Model Canvas

Print rights: Not necessary

You get a feel for the business idea you are pursuing, and the more your knowledge is based on facts, the more confident your model assumptions will be.

Further questions develop the model further towards practicality. It is helpful, for example, to document the sales roadmap by answering the following questions:

- Who are the stakeholders involved in the customer's purchase process?
- Who typically plays the role of influencer and economic buyer?
- What is the length of the sales cycle?
- What are the phases within the sales cycle?
- What is the profile of the typical buyer?
- What is the best sales strategy?

Many questions! But questions force reflection. And if you dare to think beyond your own horizon and allow critical questions, then approaches of unanticipated magnitude are often revealed.

▶   As an entrepreneur or aspiring founder, you should always think through the business model canvas and find the answers for the segments. In an existing company, this model will always be applied when you venture into a new market. You may be able to do without the process of this model if you represent an established company that "only" improves existing processes (and, for example, digitises them).

4.10   Individualized Blockchain Model

You have now illuminated your business idea from many different perspectives. Therefore, it is now time to develop a paper-based blockchain model at least as a basis for your prototype.

4.10.1 Blockchain Framework Conditions

Here you set the framework conditions that you need for your model (see Fig. 4.29).

You can see that it is about the framework in which your model is to move. Since it plays a big role for which consensus mechanism and which type of blockchain you have decided for which reasons, your preliminary assessments are now important. The stakeholders that you determined right at the beginning taking into account their influence can now become nodes in the digital blockchain environment that can validate the transactions.

Also your value promise must find its way into the model. To make this clearer, you have worked on the business model canvas and thus determined your so-called value proposition.

Which contractual modalities influence you and your processes?

These points are to be transferred to the digital world. Which new processes can be derived from the use case, which stakeholder will take on which roles in relation to the consensus? Can the value promise be translated into a mathematically accessible construct so that digital data is created? In the digital world, contracts become transactions that are processed via the chosen blockchain.

Fill in the table in Fig. 4.30 with your findings, regardless of whether you proceed top-down (from top to bottom) or bottom-up (from bottom to top) (see Fig. 4.30).

Fig. 4.29 Blockchain Framework

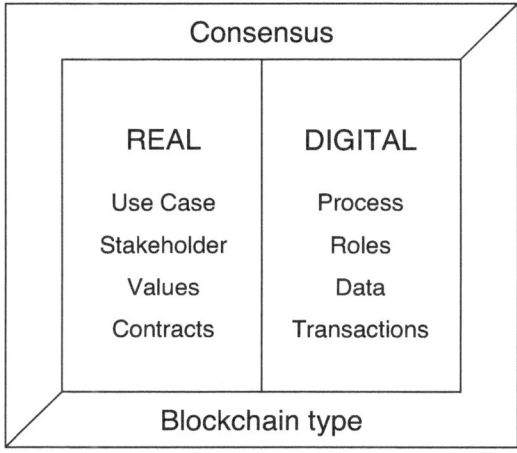

USE CASE				
Roll	Roll	Roll	Value	Value
Rights & Duties	Rights & Duties	Rights & Duties	Transfer	Transfer
Incentive	Incentive	Incentive	Data type	Data type
Census				
Blockchain type				

Fig. 4.30 Blockchain Map

You will probably get stuck when assigning roles and incentives. Here you determine which participants have which access rights and above all how you motivate these participants of your network to contribute—even if only to use the solution you have invented. For example, you have roles like "User", that is, simply someone who only uses the network without being part of it. Example: You can transfer Bitcoin from your account to another on the Bitcoin blockchain without being a miner yourself. You only need your wallet with the asymmetric key pairs to be able to carry out the transaction. Another role can be the miner (not necessarily necessary, depending on the type of blockchain selected). What task should, for example, a miner have in your system? Only create blocks or also validate transactions? Validating the blocks could be another role. You can assign this role to participants in your network. What requirements should these validators meet? Set it. And it may also be that you want to use a centrally superior administrator, which then already removes you from the classic approach of a blockchain. But maybe the main focus of your solution is traceability and thus the verification of authenticity in order to prevent fraud. Then it leads you to the approach of a private or consortium blockchain solution.[36]

---

[36] Within the blockchain community, there is something like a "directional dispute": Bitcoiners of the first hour will always speak out for a public blockchain as the only true one because it allows participation of all. But this also makes it slow. Business-oriented blockchainers, on the other hand, prefer private or consortium blockchains as the means of choice because in this approach the models can be scaled much more easily and better.

Similarly debatable is the transfer of the value promise in transactions. Once again: How can your value promise be expressed digitally? Which algorithm is necessary to enable transactions on the basis of which data type. Is the transaction carried out on the basis of asymmetric encryption or do file formats have to be converted into a hash value, for example, a PDF document, etc.

Overall, a lot of discussion is needed within the team to create clarity for all involved parties.

Background Information
An algorithm is a step-by-step guide to solving mathematical problems. As long as a problem can be expressed mathematically, it can also be represented as an algorithm. In computer science, algorithms form the general basis for writing computer programs.

There are different ways of representing algorithms:

- In the algebraic representation, the data structure is described strictly mathematically as algebra, and the calculation methods are linking rules from algebra.
- Boolean algebra is based on the characters "0" and "1" as the basis of digital electronics.
- In the diagrammatic representation, for example, a flowchart can be used to describe the step-by-step guide.
- The so-called pseudocode representation is very similar to a computer program in its writing. The algorithm is represented as a formal language.

The extended approach model, which you can see in Fig. 4.31, extends your view from your "map".

Starting from your use case, you have recorded the most important stakeholders and their roles. You have made some initial thoughts about motivation and incentive to animate the stakeholders to participate.

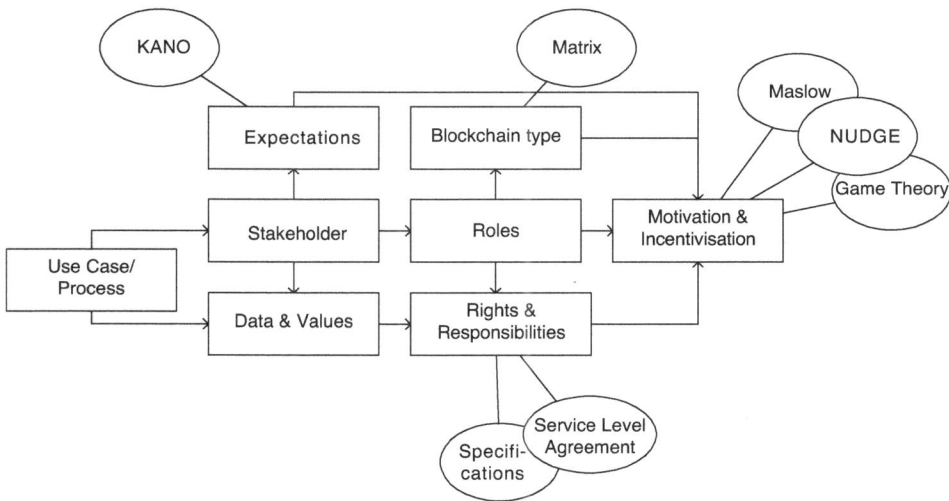

Fig. 4.31 Extended approach

This motivation to participate is described in literature (and practice) in many ways. Here are some common models that describe the interface between psychology and economics, and how this knowledge can be used for our model assumptions.

4.10.1.1 Behavioral Economics

Behavioral economics deals with the question of how to better understand economic behavior and the resulting consequences. Why does someone buy a sweater from a certain brand, take the train (and not the car), smoke, do sports, etc.?

People have to make decisions all the time. Many decisions are made consciously, but at least as many decisions are made unconsciously.

The economist Adam Smith already spoke in his work of the "invisible hand" in 1776, which leads every economic participant to pursue his/her economic interests in such a way that his/her needs are met in the best possible way taking into account the social interest in optimal resource provision. Smith explains that the self-interest of the individual leads to the welfare of a community.[37]

The behavioral economics only came back into the spotlight in the 1960s when the rational behavior pattern of Homo oeconomicus was increasingly called into question. How do people come to decisions when they have to decide under uncertainty?

Tversky and Kahneman describe this problem in their article: "Much has been discussed about the definition of rationality. There is general agreement that rational decisions must meet some basic requirements of consistency and coherence. In this article we describe decision problems in which people systematically violate the requirements of consistency and coherence. We attribute these violations to the violation of psychological principles, the subjective perception of decision problems and the associated systematic false evaluation of the different options."[38] This can lead to behavior that can have harmful effects. Berk/DeMarzo refer to the work of Shefrin and Statman, who have shown that stockholders cannot divest themselves of their stocks despite possible high losses because this would entail admitting that they made a mistake.[39]

You cannot exclude human thinking from your assumptions and approach. However, taking into account human weaknesses, you can structure your incentive system in a target group-oriented manner.

4.10.1.2 Game Theory

Game theory also provides answers to human behavior and investigates decision-making taking into account the reaction of other participants (in the game, in the respective context). A key finding is that the renunciation of options can make a participant more

---

[37] Smith (2009), pp. 271 ff.

[38] Tversky/Kahneman (1981), p. 453.

[39] Berk/DeMarzo (2019), p. 454.

successful.[40] It must be said, however, that game theory, which had its beginnings in the 1920s of the last century, still assumes a sufficient rationality of the participants.

Johann von Neumann is considered the founder of the mathematical model that describes the modeling of interactive phenomena.[41]

Blockchain uses this approach in the form of Byzantine fault tolerance. In information technology, the "Byzantine problem" describes the error that a system can behave incorrectly or completely unexpectedly.[42] It is about decisions and assessments of behavior under uncertainty with incomplete information.

You could use this approach to make assumptions about the motivation to participate in your model. The better you assess the motivation, the more likely it is to set the right incentives so that the stakeholders participate in the desired way.

4.10.1.3 Maslow's Hierarchy of Needs

Another approach that can be used to explain motivation is Maslow's Hierarchy of Needs. Maslow argues that, in addition to the incentive of "money", there are other reasons to participate in economic life. The motivation is replaced by the term "need". The pyramid structure, on the other hand, points to the hierarchical structure. Only when the first level of needs (hunger, thirst, etc.) is satisfied, the next higher level is activated. The last stage of self-realization can only be achieved if the previous stages have been successfully completed.

Within your own concept, you can also consider how to involve the stakeholders. At what stage and with which underlying needs are the corresponding stakeholders? From this knowledge, the appropriate incentives for participation can be formulated. Even if industrial society as a whole is more on levels 3 and 4 (that is, social needs and appreciation), the knowledge of this categorization helps to develop the appropriate incentive systems.

4.10.1.4 Nudge

Thaler and Sunstein, on the other hand, want to "nudge" people in the right direction so that they can make better decisions for themselves.

They almost formulate it in a heretical way that the "Homo oeconomicus thinks like Albert Einstein, stores information like IBM's supercomputer Big Blue, and has the willpower of Mahatma Gandhi."[43] The vast majority of people I know definitely do not have all three of these properties at once. And everyone I know is happy to be influenced by their irrational side. Just think about the irrational power and influence of advertising!

---

[40] Blanchard/Illing (2006), p. 704.

[41] von Neumann (1928).

[42] The problem was first raised by Akkoyunly/Ekanadham and Huber in 1975, described in detail by Lamport/Shostak/Pease in 1982.

[43] Thaler/Sunstein (2015), p. 16.

Everyone is influenceable—even to make good decisions. In their book, Thaler and Sunstein explain how people can be influenced in their decision-making process using suggestions, tips, or changes to the surrounding information.

To use this in the context of blockchain, you need to understand the motives of your stakeholders. Why do your stakeholders react to certain incentives, and are there ways to influence decisions? The nudge approach is used without "punishment", coercion, or bans.

If you look again at the extended procedure, it is clear that there are interactions between the stakeholders and the values derived from the use case. If there is a value promise from which data and transactions can be derived, then this always goes hand in hand with expectation management.

Users of a blockchain solution you have designed have expectations of all kinds. It is important to know these expectations and to meet them.

The so-called KANO model supports you in finding out what a user/customer/stakeholder expects from your offer.

4.10.1.5 KANO Model

This model can be assigned to quality management, because it describes the relationship between customer satisfaction and customer requirements (cf. Fig. 4.32).

The model consists of two axes. The y-axis represents customer satisfaction from highly satisfied (top) to disappointed (bottom), while the x-axis reflects the realized quality characteristics, i.e. how well the product is perceived (far right) or how badly (far left).

The indifferent zone around the intersection says that the customer is still unable to assess whether he/she likes the product or not.

The curve in the upper quadrant describes the enthusiasm requirements, the straight line represents the quality and performance requirements, and the curve in the lower quadrant is the representative of the basic and fundamental requirements.

Basic requirements are therefore so basic and self-evident that they only become aware of them when they are not met (implicit expectations). If these basic requirements are not met, dissatisfaction arises. If they are met, however, this does not automatically lead to satisfaction. The increase in benefits compared to the differentiation from competitors is very low.

The performance features eliminate customer dissatisfaction. These features create customer satisfaction and are aware of the customer. That is, the more of these performance features are fulfilled, the more satisfied the customer is. It is, as the line also shows, a proportional increase.

The enthusiasm features, on the other hand, are benefit-providing features that the customer does not necessarily expect. These features are not self-evident. They distinguish the product from the competition and evoke enthusiasm. A small performance increase can lead to an disproportionate benefit. The differentiations from the competition can be small, but the benefit enormous. And here too, the more, the better.

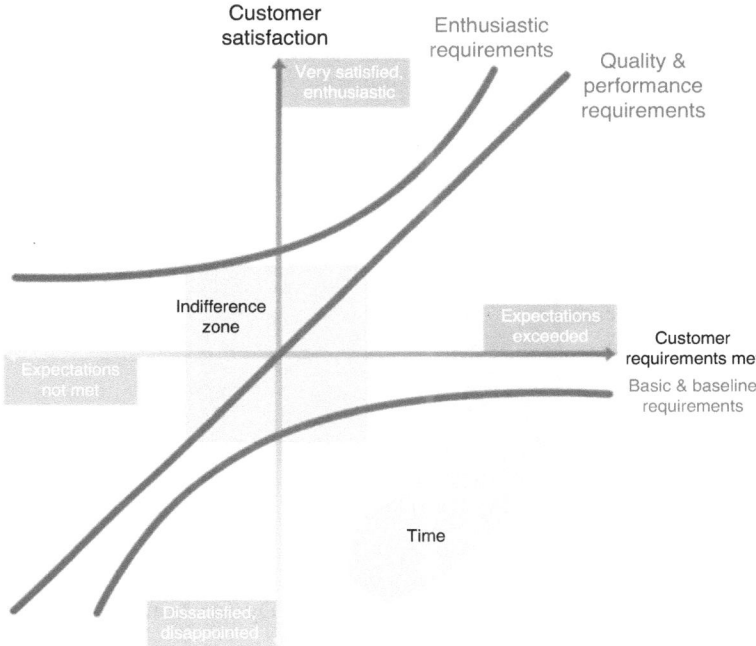

Fig. 4.32 KANO model. (From Ronald Jochem "What does quality cost"; with kind permission © Carl Hanser Verlag Munich 2019 All Rights Reserved)

However, not every customer has the same expectations, and not every customer is influenced by the same features. Priorities are set differently. In addition, there is a dynamics in the features, i.e. what is experienced today as an enthusiasm feature can become standard tomorrow and even "slip" into a basic feature.[44]

This model has two more features: on the one hand so-called insignificant features, which are of no importance to the customer in the event of their presence or absence. They can therefore neither create satisfaction nor dissatisfaction. On the other hand, there are then so-called rejection features that, in the event of their presence, lead to dissatisfaction, but in the event of their absence do not lead to satisfaction.

In order to find out how these features are to be sorted from the customer's point of view, the inventor of the model has provided for a survey with positively or negatively formulated questions:

---

[44] As an example, an iPhone from Apple is mentioned here. The touchscreen was something completely new at the time of introduction and caused enthusiasm. Today all smartphones have a touchscreen, so it has become a basic requirement.

- Functional—that is, positively formulated: What would you say if the product has "xyz"?
- Dysfunctional—that is, negatively formulated: What would you say if the product does not have "xyz"?

Kano sets out possible answers for both questions in the following form in order to derive an classification of the features:

The possible answers are:
- "I would be very pleased"
- "I take that for granted"
- "I don't care"
- "I can still accept that"
- "That would bother me a lot"

Depending on whether you have formulated the question positively or negatively, the following features result:

- Basic features => Functional "I assume that" & Dysfunctional "That would bother me"
- Performance feature → functional "That would make me very happy" & dysfunctional "That would bother me very much"
- Enthusiasm feature → functional "That would make me very happy" & dysfunctional "That doesn't matter to me"
- Insignificant feature → functional "That doesn't matter to me" & dysfunctional "That doesn't matter to me"
- Rejection feature → functional "That would bother me very much" & dysfunctional "I assume that"

The interpretation of this model can be queried via customer tests and/or customer interviews. With a sufficiently large number of testers, you will receive good information about your prototypes as well as existing products.

All the findings that can be gained from this approach can be used very well to make the incentive and motivation of the participants so unique that the corresponding stakeholders are actually willing to participate in the roles and tasks provided by you. Depending on the intensity, you will design the block type to be used.

Finally, we can still consider how data and values can be transferred to a requirements specification or as characteristics in service level agreements. You need them to explain your expectations to the developer.

A software requirements specification lists the requirements you have for implementing this project. They are described binding in the document. This includes, in addition to functional and non-functional system requirements, guidelines for the technical framework and other resulting specifications. Delivery conditions and acceptance criteria should also be listed. A structure for describing the requirements could look like this:

- Objective: This is where the description of the goal to be achieved by using the product is made.
- Product Use: Here you state for which target groups the product is intended and what requirements it must meet.
- Product functions: The core functions are formulated and described from the client's perspective here.
- Product data: The data to be stored is specified here.
- Product services: Time, data volume and accuracy in relation to the main functions are specified here.
- Quality requirements: The most important quality requirements such as reliability, usability, efficiency are specified here.
- Under "Additions", all special features that go beyond what has been specified so far can be recorded (e.g. extended requirements for the user interface, etc.).

In addition to the requirements specification, a service level agreement is also a way to make agreements with a provider.

Further details on the requirements specification and the service level agreement can be found in Sect. 2.10.

4.10.2 Creating the First Prototype

After you have completed all the exercises, it is almost time for the "main event". You are supposed to use a "flowchart"[45] created by you to represent the process.

▷   Please only work on one process at a time at the beginning. It is advisable not to get too complex, otherwise you might overlook details. This leads to problems in the downstream processes. In addition to cost explosions, the time frame can also be blown up, and in the worst case you will not be able to bring your new application to market on time despite the announcement.

The elements of a blockchain (transactions, nodes, ledger, user, smart contracts) not only stand for themselves, but also fulfill certain tasks. Choose the ones you need for the first

---

[45] It is the predecessor of a "real" technical flowchart. However, this description will help you now, at the end of all the exercises, to take up the common understanding within the group and transfer it into this graphical representation.

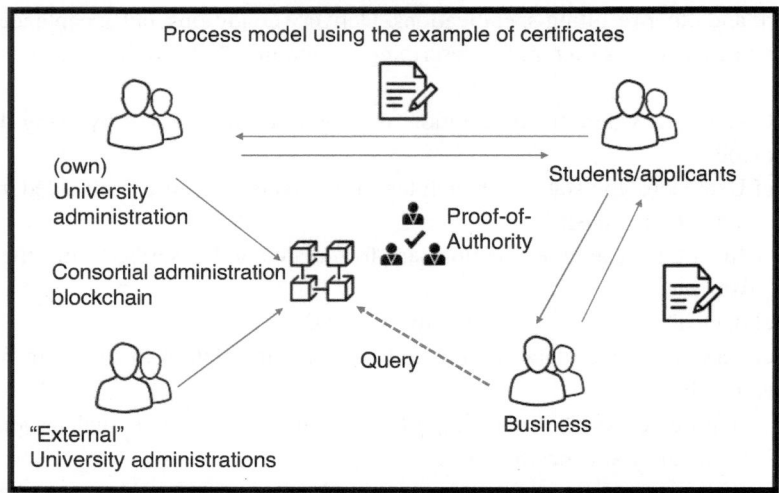

Fig. 4.33 Process mapping on the example of university certificates

(possibly also rough) representation from the attached symbol list. With the help of the glossary, you can assign the corresponding symbols their roles and tasks.[46]

Please proceed as follows:

- Assign the interest groups you have included on the map.
- Which roles/tasks should your stakeholders assume within your blockchain solution? Briefly describe which input is expected from which stakeholder.
- Decide which stakeholders will "host" the blockchain and validate the transactions.
- Mark the data flow.
- Describe your blockchain model in words to make sure the group has the same understanding.

As an example, you can see in the following graphic (cf. Fig. 4.33) how such a representation can look like on the example of university and certificate creation (in the course of the value-added analysis it has become apparent that the potential for certificate fraud is a so-called pain point for all universities).

As you can see from the figure, the approach of the blockchain is a so-called consortium. The decision against a public blockchain is based on the fact that it will "only" be about this one task, namely to store certificates securely. The more administrative acts are carried out via blockchain solutions, the sooner the consolidation towards a public, permissionless variant will take place.

---

[46] Only a few symbols are shown as icons in the template to at least show the first frame. Of course, you can always define more icons and integrate them into your system.

Back to the current approach: Here, the Proof of Authority has been chosen as the consensus mechanism (see Sect. 2.4). The operators of the blockchain are therefore the universities, which determine the universities as validators via the university administrations (it is possible that initially only a limited number of universities will play the role because one does not get every university involved at the beginning). These universities do not have to hold a stake to create and validate blocks. They are credible by office and can thus ensure the smooth running of this blockchain.

Users of the system are, in addition to the universities (double role), the students, applicants for study places and employers.

Applicants for study places apply online twice a year for the places made available by the university. To do this, an interested party goes to the start page of the "Digital-oriented Service Procedures" (DoSV) and finds out the necessary details in order to apply.[47] At a university participating in this procedure, the application requests come from around mid-November to mid-January (for the respective summer semester) and from around mid-May to mid-July for the winter semester. Although the application process appears to be digitized, even if the applicant submits all his/her previous evidence via the DoSV platform—the screening process at the universities is manual. Specially trained employees of the administration check the incoming documents for credibility and, in case of success, send the admission to the applicant. (Whether he/she accepts it or rather goes to another university should not be pursued in this process section). For the sake of simplicity, this process is not pursued any further in this description.

At the end of their studies, students receive their final transcript.[48] If you only take the process of creating this certificate, it becomes obvious how much manual work is still involved in a process that is actually automatable. In addition, there is no guarantee that an dissatisfied student will optimize his/her results. There have been plenty of spectacular certificate frauds in the past. This is where the example comes in. A university issues the certificate of graduation for the graduate. The content of the certificate is stored as a hash value on a consortium-operated blockchain. Although the graduate receives his/her certificate and a certificate, the content of the certificate has become verifiable and verifiable through the storage on the blockchain. If the graduate applies for a job with his/her certificate, the potential employer can check the (hopefully) attached hash value to see if the document is real. Such a verification process can look like this from the perspective of the potential employer:

---

[47] https://sv.hochschulstart.de/index.php?id=8, Accessed: 03. October 2019.

[48] Students also have the opportunity to have their semester certificates issued for the courses completed during their studies. This is always necessary, for example, when students apply for an internship because it is part of their studies. There are also ways to overcome the existing system and make individualized adjustments. This part of the process is again left out to keep the schema as simple as possible at first.

- Inquiry whether this certificate exists on the "university blockchain".
- Check hash values: If the hash values match, no changes have been made to the document. For this purpose, the potential employer will convert the certificate available to him into this value using a predefined hash generator and then check whether his/her value matches the one on the "university blockchain".

The consequences of such a solution are manifold.

The company can be sure that the applicant has actually achieved the academic performance that is stated on his/her certificate.

Universities benefit in two ways: On the one hand, the certificates become very valuable because they can be checked for authenticity, and on the other hand, a real relief for employees in administration can be achieved through the consistent automation of the (entire) process.

For companies, the authenticity check is also significant because they can be sure to actually have the candidate with the "real" performance promise in front of them.

4.10.3 Further Information

The possibilities of representation are manifold, and surely you have further ideas how you want to proceed graphically (cf. Fig. 4.34).

Conformance testing scenarios are used in process models to define the permissible behavior of a process. A key prerequisite for the test is that the events in event traces can be related to the activities of a process model. For example, in an event trace t (time) = < Event 1, Event 2, … Event n > and the process model shown in the figure, the events in t must be mapped to activities in the model. Otherwise, it is impossible to understand which activities took place in reality and whether t matches M or not. There is also freely available test software for this.

All of these representations serve to illustrate your analyzed process and to find bottlenecks. When testing your processes, you can consider where your smart contract makes sense.

4.10.3.1 Smart Contract

Smart contracts are software-based computer programs that can map or verify contracts (see Sect. 3.2).

Please look at your identified processes and question how smart contracts can be used meaningfully. By using smart contracts, you can not only create event-based "if-then-queries", but you can also formulate and program an event to transfer it to the blockchain. Smart contracts have other—partly for an understanding grown from the analog world—unusual properties. For example, smart contracts can capture a greater amount of

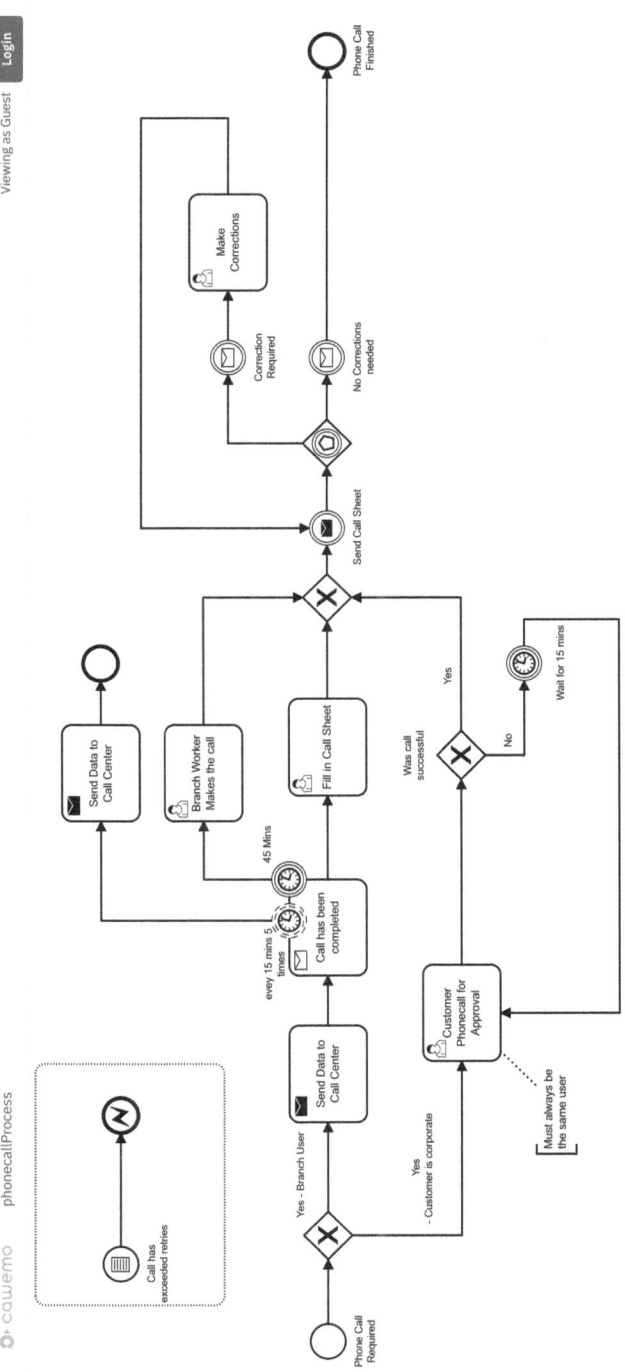

Fig. 4.34 Process representation. (with kind permission from © Camunda 2020 All Rights Reserved)

information. In addition, they are dynamic, i.e. they can transfer information and make certain decisions. This distinguishes them from their paper predecessor.[49]

▶   In your team, put together a list of what your smart contract should ideally be able to do with all its facets in the process. This list will then be part of what you clarify with the developers you have commissioned.

Background Information
As a classical example, leasing a car is often mentioned in this context. As early as 2016, this was the first application case between Visa and DokuSign. Accordingly, after making a decision for a car and the leasing contract, a potential customer will make extended options such as e.g. number of kilometers to be driven per year etc. These details as well as payment conditions are recorded in the leasing contract, which is then encoded as a smart contract and stored on a blockchain. If the customer now does not pay the agreed leasing rate on time or punctually, this is an event that triggers the "Action" according to the smart contract logic. First, this may be a reminder. If the rate does not arrive at the leasing company within the grace period, the smart contract locks the vehicle.

Thus, the car becomes a "smart device" that interacts with the environment through sensors. More scenarios are conceivable: Imagine that the driver of the leasing vehicle disregards the traffic rules with his/her driving style. An automated speed control could then come into effect. If it is an electric car, the electricity to be drawn could be directly debited from the driver's bank account when the driver refuels his/her electricity. If the driver adheres to the maintenance intervals and the contract workshop, there could also be a reward for this via the smart contract.

As you can see, the imagination knows no bounds.

After you have described your requirements for the smart contract, the question arises on the one hand which effects these have on your so-called front-end (that is what a user of your potential application sees). On the other hand, you have to take into account the back-end in which your blockchain runs. These two levels must be connected to each other in order to enable the corresponding communication of the two levels.

4.10.3.2 Front-End

Front-end software components are those that visualize and provide the user with a user interface for the underlying database/server (thus the link to the back-end).[50] In this context, one also speaks of a GUI (graphical User Interface). This graphical User Interface makes the application software operable by means of graphical symbols or control elements. (Either this can be done via a mouse as a control device, or, as with smartphones, by touch).

In order to make this user-friendly, free software can be used to create so-called mock-up versions with which the navigation in the app can be simulated. Just enter

[49] Tapscott/Tapscott (2016), p. 101.
[50] Kersken (2019), p. 208.

"mockup generator for free" in your browser, and you will get a variety of offers. This way you can test whether your assumptions can be operated intuitively by a user. You get a feel for whether the sequence you have chosen makes sense or whether it "hangs". You can also test designs.

▶ If you have determined which of your suitable processes is to be offered as a new product and you need a user-friendly interface for this, it is recommended that you first only design your idea on the free portals. The final implementation is better entrusted to a web designer and/or a front-end programmer.

The frontend is to be understood as a graphical user interface that allows the user to access a (stored in the backend) database.

4.10.3.3 Back-End

While Front-End is what the user sees on his/her screen, the back-end is so to speak the engine room, in which "everything" has to happen and work. Data processing and data storage take place here, based on the underlying business logic. If you take the "client-server structure" mentioned in Sect. 1.3 , then the back-end refers to the server side. This generally includes a web server that communicates with the database to serve the request presented by the front-end (see Fig. 4.35).

File:

As can be seen from the figure, the user can call up the calendar function in the front-end. For this purpose, this application is connected to the back-end. In the back-end, the various servers that perform different tasks can be seen.

It can thus be said that front- and back-end are the two sides of a coin.

4.11   First Requirements for the Developer Formulate

You have gone through all levels of this procedure, have certainly discussed a lot in the team and also rejected. You have certainly also noticed in the discussions that the understanding of one or the other thing is partly very different. This may be because you have put together an interdisciplinary team whose participants also bring their own perspectives and their own understanding of certain terms according to their task. This makes it clear how difficult it is for a programmer to program the digital product for you that you would like.

A too holistic approach will cause dissatisfaction on both sides. Therefore, you have taken the trouble and gone through so many different levels in order to crystallize out more and more clearly where your potential for renewal lies. Now you have to "process" this knowledge you have collected for your programmers.

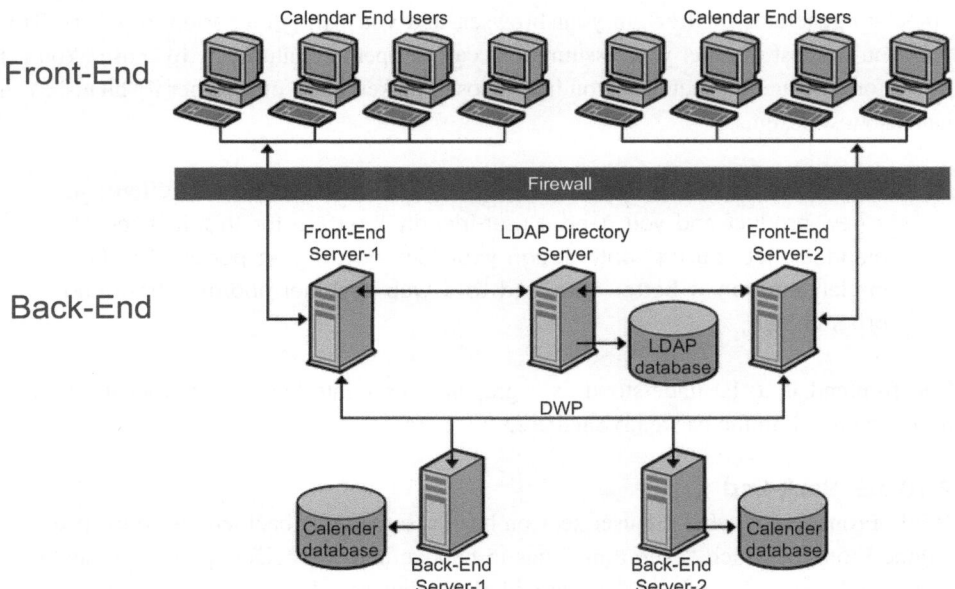

Fig. 4.35 Relationship between Front- und Back-End. (mit freundlicher Genehmigung von © Oracle 2010 All Rights Reserved)

Software development is a complex process. For non-software developers, the problem is to understand how far the project is implemented, as software is an initially invisible product.

It can be helpful to describe the process you have filtered out as a flowchart with the "if-then-logic" you desire. You can use this as a first basis in conversation to clarify with the developer and, if necessary, his/her team how your ideas can be implemented in software code. Agree with the development team on project-specific structures and scripts that can be checked against a given metric.[51] Please introduce an error and change management from the beginning. Errors that are only recognized too late in the course of the project not only cause an disproportionate amount of effort to remedy this problem, but also extreme costs. As early as 2001, scientists Barry Boehm and Victor R. Basilie published a still valid Top-10 list on the topic of software errors and their effects.[52] Top 2 of this list states that 40–50% of the effort is caused by avoidable rework.

This avoidable rework can be reduced by using Scrum tools. Define which components belong to the product and how these components are recorded. You can manage

---

[51] Popp (2013), pp. 60 ff.
[52] Boehm/Basili (2001).

Table 4.4 Documentation requirements according to Popp

Element	Useful additional documentation
Requirements document (e.g. Use Case); Architecture and design document User documentation	• Source and description of the document template used • Reference to sample documents that demonstrate proper use of the template • Guidelines for creating diagrams. This includes both the tools to be used and the instructions for inserting graphics into the documents
Build script	• General description of the build process • Documentation of the requirements for running the script • Description of the possible command line parameters
Meta- and configuration data	• Description of who is allowed to change the data and when and how (The contents of the files should not be described—this is either part of the user or the operational documentation)
Source code	• Coding standards • Guidelines for formatting. If tools for automatic formatting are used, the configuration and use of the tools should be described • Guidelines for documenting the source code
Tool	• Used version and source of reference, preferably in the form of links to the installation package • Installation and configuration instructions

these documents and data via a so-called repository (a directory for describing and storing digital objects as a software library).

According to Popp, the elements summarized in Table 4.4, again will this table be inserted even though I can't see it here? which are necessary for the development of software, belong in the directory.[53]

This documentation allows you as a non-developer to better understand what work developers have to do and what has already been done.

4.12 Conclusion

The creation of "event-to-activity" maps is extremely complex and currently not machine-implementable.

You have become familiar with so many different models and considered various aspects from different perspectives. All of these different approaches allow you to see the weaknesses in your processes. And at this stage you have (working) time of the team, pens and paper consumed. All in all, still manageable resources.

---

[53] Popp (2013), p. 31.

Unfortunately, it is a great challenge to establish a correct assignment between events and activities. Existing techniques that take on this challenge can therefore not provide a definitive solution, but a sufficiently good trend. The reason why mapping techniques do not provide definitive solutions is that the information they can take into account when constructing mappings is often not sufficient to identify relationships with certainty.[54]

For this reason, we have gone through various mapping models that deal with the question of whether your idea makes sense for a blockchain application. By looking at things from different angles, you have to deal with different questions. This presents a great opportunity because the questions will help you think more openly and "outside the box."

With your solution approach and the accompanying documentation, you can now aim for the next level. You have created flowcharts to conjugate the process flows, you need to describe how, in your opinion, the blockchain technology that is anchored in the background will be connected to the so-called front end. And you have thought about how a user can use your product.

Once again: the task is great. Therefore, as explained in Scrum, it is advisable to define small sub-goals that need to be achieved. This way you can much more quickly recognize when the project needs additional iterations or even new beginnings for partial tasks.

And in many cases it is not about reinventing the wheel. The challenge is to rethink what already exists and to allow for different perspectives.

You should therefore ask yourself whether the conditions for the introduction of a blockchain solution are met. Which blockchain do you want to use? A public one that is open to everyone or, at the other extreme, a private one that is only accessible to selected participants? What added value does the use of a distributed (ledger) system provide? Both centralized and distributed systems have their own advantages and disadvantages.[55]

What is the benefit of the application you have found? Consider that your idea and also the blockchain system should create value for its users. (Technical note: It should be noted that even a very sophisticated system architecture can never compensate for a weak or poor application idea.)

How are the nodes compensated for the resources they provide and the maintenance of their integrity? The integrity of the blockchain is ensured by the use of incentive systems based on fee income and work certificates. The knowledge and understanding of such compensation to ensure integrity is a key aspect of the analysis (see Consensus Mechanisms Chap. 2)

---

[54] Baier et al. (2014), pp. 127 ff.

[55] The centralized systems are not naturally deficient, but rather choose a different architectural concept that provides good service to a large number of applications and can be maintained in this way.

Does the consolidation/anchoring of a blockchain depend on the existence of a community of developers? Certain projects that support the development of a new blockchain depend on the support of an open community that supports them for their implementation. Its existence, number and activity are relevant elements for assessing its viability. Therefore, you should consider which blockchain provider can offer you the "framework" you need while at the same time checking how large and secure the community is.

In projects where the solution includes the introduction of an Initial Coin Offering/Security Token Offering, it is necessary to check whether its use is really necessary or more than a source of financing for the company or has a predominantly speculative character. For more information, see Chap. 6.

In which phase is the business idea you have developed in the context of the entire business development? Currently, many projects are only proposed on the basis of an idea; other ideas are supported by a proof of concept, and others may not be market-ready yet, but at least have access to the market.

In this context, you can check your marketing and marketing activities. How does the new concept fit into your previous activities and the overall company? How can the marketing tools be used for the new business idea you have developed to make the goals of this project known (see also Sect. 4.8)

You have spent a lot of time with and in your team to ask all these questions and models. The purpose is, in addition to good arguments for or against a blockchain-based solution, to determine whether the business idea is valid enough to justify the further effort (of a blockchain implementation).

References

Baier T, Mendling J, Weske M (2014) Bridging abstraction layers in process mining. Inf Syst 46:123–139

Berk J, DeMarzo P (2019) Grundlagen der Finanzwirtschaft, Analyse, Entscheidung und Umsetzung, 4., akt. Aufl. Pearson, Hallbergmoos

Blanchard O, Illing G (2006) Makroökonomie, 4., akt. u. erw. Aufl. Pearson, München

Boehm B, Basili VR (2001) Software defect reduction top-10-list. IEEE Comput 34(1):135–137

Dillerup R, Stoi R (2013) Unternehmensführung, 4. Aufl. Vahlen, München

Duden (1990) Das Fremdwörterbuch, Bd 5, Drosdowski G et al (Hrsg), 5., neu bearb. u. erw. Aufl. Bibliografisches Institut AG/Dudenverlag, Mannheim

Finyard. https://www.finyear.com/Blockchain-A-legacy-of-transparency_a36758.html. Accessed: 18. Sept. 2019

Gardner J (2017) Do you really need a blockchain? Distributed Magazine, Nashville

Gerzema J, Lebar E (eds) (2008) The brand bubble, the looming crisis in brand value and how to avoid it. Jossey-Bass, San Francisco

Golem Plattform (2019) https://www.golem.de/specials/zynga/. Accessed: 12. Sept. 2019

Hochschulstart (2019) https://sv.hochschulstart.de/index.php?id=8. Accessed: 3. Oct. 2019

Jochem R (2019) Was kostet Qualität? 2., überarb. Aufl. Hanser, München

Kersken, S (2019) IT-Handbuch für Fachinformatiker, der Ausbildungsbegleiter, 9., erw. Aufl. Rheinwerk, Bonn

Kletti J, Schumacher J (2014) Die perfekte Produktion, Manufacturing Excellence durch Short Interval Technology (SIT), 2. Aufl. Springer Vieweg, Berlin

Koens T, Poll E (2018) What blockchain alternative do you need? https://www.cs.ru.nl/E.Poll/papers/blockchain-alternative2018.pdf. Accessed: 15. Sept. 2019

Koubek A (Hrsg) (2015) Praxisbuch ISO 9001: 2015, Die neuen Anforderungen verstehen und umsetzen, 3. Aufl. Hanser, München

Kühnapfel JB (2019) Nutzwertanalysen in Marketing und Vertrieb, 2. Aufl. Springer Gabler, Wiesbaden

Meunier S (2016) When do you need a blockchain, Medium. https://medium.com/@sbmeunier/when-do-you-need-blockchain-decision-models-a5c40e7c9ba1. Accessed: 8. Nov. 2019

Mintzberg H (1983) Power in and around organisations (The theory of management policy). Prentice Hall, Englewood Cliffs

Mittelmann A (2021) https://www.artm-friends.at/am/km/WM-Methoden/WM-Methoden-285.htm. Accessed: 1. May 2021

Mulligan C (2018) Blockchain beyond the hype. https://www.weforum.org/agenda/2018/04/questions-blockchain-toolkit-right-for-business

Neumann J von (1928) Zur Theorie der Gesellschaftsspiele. Mathematische Analen 100:295–325

NISTIR (eds) (2018) Yaga, Dylan/Well, Peter: blockchain technology overview. https://doi.org/10.6028/NIST.IR.8202. Accessed: 8. Nov. 2019

OECD (2019) Bildung auf einen Blick 2019, OECD-Indikatoren. wby Media. Bielefeld

Oracle (2010) Multiple front-end servers with multiple back-end servers, Sun Java System Calendar Server 6 2005Q4 Administration Guide

Osterwalder P (2004) Business Model Generation. Ein Handbuch für Visionäre, Spielveränderer und Herausforderer. Campus, Frankfurt

Pastoski S (2004) Messung der Dienstleistungsqualität in komplexen Marktstrukturen. Deutscher Universitätsverlag, Wiesbaden

Peck ME (2017) Do you need a blockchain. https://spectrum.ieee.org/computing/networks/do-you-need-a-blockchain. Accessed: 8. Nov. 2019

Pfeffer M (2014) Bewertung von Wertströmen, Kosten-Nutzen-Betrachtung von Optimierungsszenarien. Springer Gabler, Wiesbaden

Popp G (2013) Konfigurationsmanagement mit Subversion, Maven und Redmine, Grundlagen für Softwarearchitekten und Entwickler, 4., akt. u. erw. Aufl. dpunkt, Heidelberg

Pricken M (2007) kribbeln im Kopf, Kreativitätstechniken & Denkstrategien für Werbung, Marketing & Medien, 10., komplett überarb. Aufl. Hermann Schmidt, Mainz

Smith A (1776/2009) Wohlstand der Nationen. Anaconda, Köln

Suichies B (2016) When do you need a Blockchain? (In: Meunier, Sebastian 2016). https://medium.com/@sbmeunier/when-do-you-need-blockchain-decision-models-a5c40e7c9ba1. Accessed: 8. Nov. 2019

Tapscott D, Tapscott A (2016) Blockchain revolution, how the technology behind bitcoin is changing money, business, and the world. Penguin Random House, New York

Thaler RH, Sunstein CR (2015) Nudge, Wie man kluge Entscheidungen anstößt, 5. Aufl. Ullstein Econ, Berlin

Thommen JP, Achleitner A-K, Gilbert DU, Hachmeister D, Kaiser G (2017) Allgemeine Betriebswirtschaftslehre, Umfassende Einführung aus managementorientierter Sicht, 8. Aufl. Springer Gaber, Wiesbaden

Tversky A, Kahneman D (1981) The framing of decision and the psychology of choice, vol. 211. Science Magazine, Nr. 4481

Weidner GE (2017) Qualitätsmanagement, Kompaktes Wissen, Konkrete Umsetzung, Praktische Arbeitshilfen, 2., überarb. Aufl. Hanser, München

Werner J, Mandel P, Zarnekow R (2020) In: Fill H-G, Meier A (Hrsg) Blockchain, Grundlagen, Anwendungsszenarien und Nutzungspotenziale. Springer Vieweg, Edition HMD, Wiesbaden

Wüst K, Gervais A (2017) Do you need a blockchain? https://eprint.iacr.org/2017/375.pdf. Accessed: 8. Nov. 2019

Yaga D, Mell P, Roby N, Scarfone K (2018) Blockchain technology overview. National Institute of Standard and Technology, Nistir 8202. https://doi.org/10.6028/NIST.IR8202. Accessed: 9. Nov. 2019

The Code is the Law

Abstract

Soft- and hardware regulate the architecture underlying the internet. In the digital world, a piece of software determines the course of things. It is not surprising that the impression can arise that a program code defines conditions that are considered inviolable. Blockchain technology as one facet of digitalization cannot escape this approach. And yet there is more to technology in the broader sense than "just" writing program lines. Therefore, this chapter reflects on the statement what "The code is the law" means in the context of blockchain technology.

This imperative must be reconsidered. If it is translated with Kant's categorical imperative[1] then the question arises as to how programming should "behave" correctly. Kant's imperative is the attempt to find a standard for just action—but can programming do this at all?[2]

In addition, however, it must be taken into account that neither the Internet nor a blockchain technology act in a lawless space.

As early as 2006, Lessig described in the preface to his book "Code" the development from the 1st to the 2nd edition in a nutshell to the effect that, around the turn of the millennium, it was still assumed that cyberspace was a lawless space in which online

---

[1] Kant's categorical imperative: "Act in such a way that the maxim of your will could always at the same time serve as the principle of a general legislation."

[2] The exciting question of ethics within and behind programming is not further illuminated here. However, it should definitely be included in the planning of new software solutions, as the sensitivity is increasing worldwide.

© The Author(s), under exclusive license to Springer-Verlag GmbH, DE, part of Springer Nature 2022
K. Adam, *Blockchain Technology for Business Processes*,
https://doi.org/10.1007/978-3-662-65818-5_5

life was separated from offline life and governments had no way of enforcing rights and obligations. This view has changed in just a few years, and the idea that the Internet is an unregulated, lawless space has disappeared.[3]

A computer program based on one (or more) algorithms must also be measured against this. The path from influence to manipulation can be very short.

Sauerwein et al. show in this context how helpful the algorithmic selection is on the Internet.[4] Algorithmic selection, as an input-throughput-output model, is based on different relevant input data (e.g. user profile), which are used, for example, by filtering in the throughput, to display different results in the output (e.g. recommendations, based on the user profile and corresponding website visits [in this way, Amazon, among others, selects its customers and makes recommendations based on the insights gained]). Given the wealth of information available on the Internet, this form of filtering is helpful on the one hand, but on the other hand there are also considerable disadvantages. For example, this filtering can lead to various risks, e.g. distortion of reality, filter bubble, restriction of freedom of communication, etc.[5] The algorithm calculates and displays results without "its own" valuation (this would require a consciousness comparable to that of humans in machines and computers). The programming of the algorithm and the resulting filtering is done by humans.

The blockchain technology (in the meta-sense) enables both public and private actors to use the blockchain to build their own rulebook, which is implemented with self-executing, code-based systems. Here, a community of people decides as part of such a network which rules are to be observed. With blockchain technology, this can then be traced by everyone and is therefore considered trustworthy.

But the key question is, what can a code accomplish.[6] How is the code developed and what implications are associated with it? Can and may a code, which is developed by a (so far human) programmer, be equated with the law and thus become law?

Every programmer will agree to that at first, because what a programmer defines in his/her code is followed by those who use his/her program; whether it is the way we are led through an application, or how the code actually runs in the background.

There is a close interaction between risks and benefits, because if risks are to be minimized, this is usually associated with a restriction of benefits. Against this background, a blockchain-based solution appears attractive because this technology can efficiently automate interactions between technology and environment on the basis of laws and regulations. A worldwide discussion has broken out about universally applicable governance

---

[3] Lessig (2006), p. IX.

[4] Sauerwein, Florian/Just, Natascha/Latzer, Michael (2017), p. 1, accessed on October 15, 2019.

[5] For example, Eli Pariser, American political activist and blogger, who in 2011 realized that different users can receive different results for the same Internet query.

[6] Cf. De Filippi/Wright (2018), p. 195.

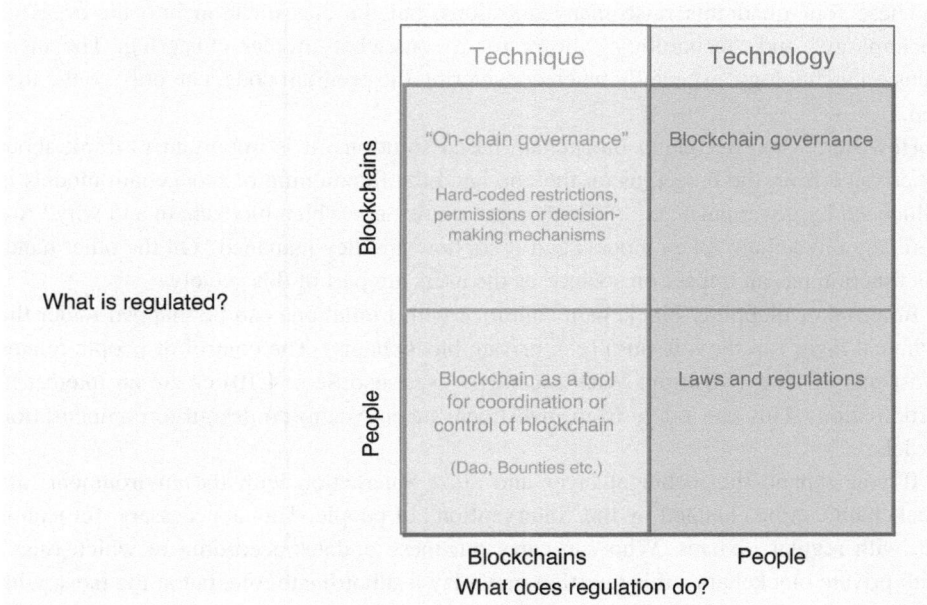

Fig. 5.1 Governance levels. (own representation based on Prewitt and McKie)

rules for the blockchain industry. Can blockchain technology improve existing IT governance?

Weill defines IT governance as follows:

> *IT governance provides the framework for decision-making rights and responsibilities to promote desired behavior in the use of IT.*[7]

In corporate management as well as in the public sector, the federal decision-making model has a tradition. This federal form of decision-making can be excellently mapped by a blockchain network with the help of the superordinate level with the relevant environmental factors and the interactions on the artificial physical systems.

This allows us to derive the scope of where and how a code works and how it should be orchestrated through governance approaches. In their Medium article, Prewitt and McKie therefore derive four quadrants in order to make an assessment (see Fig. 5.1).[8]

---

[7] Weill (2004), p. 3.

[8] Prewitt/McKnie under https://medium.com/blockchannel/blockchain-communities-and-their-emergent-governance-fdf24329551f, accessed on October 12, 2019.

These four quadrants raise many questions, but the classification into the areas of "technology" and "technology" allows for a somewhat simpler clustering. The mere focus on technology, especially when considering the program code, can only be the first step.

However, when designing blockchain-based solutions, it is important to think about governance from the outset, as on the one hand the functioning of blockchain models is influenced by governance, e.g.: Which communities use which blockchain and why? Are there any blockchain token models and if so, how are they managed? On the other hand, this function has an impact on society, as the users are part of this society.

In terms of mapping, blockchain solutions with limitations can be mapped under the technical layer (on the left side) (e.g. private blockchains). The control of people relates to aspects of incentivization: What incentives (see also Sect. 4.10) create an interest in participation? This can range from investment incentives to co-design to compensation models.

If you expand the technical layer and allow interaction with the environment, the blockchain can be changed by the "intervention" of people. This is necessary, for example, with regular updates. Who can carry out these updates according to which rules? With private blockchains, this question may play a subordinate role, but at the latest with public blockchains, this represents an intervention in the protocol including the consensus mechanism—and then it's all about the whole!

The last quadrant shows the ratio of how people, e.g. blockchain users or blockchain developers, adapt the existing legal framework in the blockchain world through superordinate institutions.

In particular, in this area, the strength of blockchain technology could show itself in the future. The maturing economic blockchain system can, with clever use of transparency and tamper-proofing as well as the automatic execution of smart contract codes, both blockchain-inherent and legislatively prescribed regulations and regulations more efficiently and safely.

All blockchains have rules that organize their processes. If blockchains evolve, they may need to change these rules. Blockchain governance thus refers to the system through which decisions about the development of the blockchain are made. A blockchain network is a complex system involving a variety of actors that you do not know and cannot trust—basically. The protocol of a blockchain should therefore ensure that every actor has an incentive to cooperate and that the costs of violating the consensus and protocol are higher than the potential gains.

Example

The following example should illustrate this: In sect. 3.2 the mode of operation of smart contracts is shown using the example of a drink vending machine. Let us use the example further.

You are thirsty and want to drink something while you are waiting for the train. The drink vending machine on the platform offers its services against cash in the form of coins. You don't have any change with you. Nevertheless, you will not crack the machine and thus violate the consensus, because the costs (compensation/prosecution) of this violation are much too high in comparison to what you receive as "profit" (namely a cooled drink). ◀

Thus, the program code with the protocol and consensus mechanism surrounding it has the ability to influence the behavior of people—similar to a law. And the individual who uses this code packaged in an application expects legal certainty.

The ability to document contractual conditions in a revision-proof manner using smart contracts allows for a variety of new applications, taking into account the governance issues mentioned above. This is accompanied by

- *Reliability:* If a smart contract has been programmed correctly, difficulties in interpreting the contract terms are virtually excluded. The loss of documents is also ruled out.
- *Security:* If smart contracts are programmed on the basis of a blockchain, they are protected against hackers by encryption methods. Nobody can change the negotiated contract terms retrospectively.
- *Efficiency:* Programming a smart contract well requires a lot of time, but far less than the corresponding bureaucratic processing. This saves time and money for contracting parties.
- *Independence:* Smart contracts enable so-called middlemen or intermediaries to be left out of the process (e.g. banks, lawyers, notaries, etc.). The only verification is the immutable storage on a blockchain. The program code of the contract decides whether the contract terms have been complied with, often on a step-by-step basis. One is tempted to say: "The code is the law".

As already explained, however, much of this is still in the making and, like other complex systems, blockchains consist of many different parts that interact with each other in sometimes unpredictable ways—and are therefore difficult to control or regulate. It might be possible to regulate the actions of each individual part. But since the whole is greater than the sum of its parts, governance cannot be achieved without a proper understanding of the different components that make up this whole and the power dynamics that exist between them.

Perhaps we should therefore follow the "imperative" "The code is the law" with the following thought process: "The code is decisive for determining the rules and regulations" (Der Code bestimmt maßgeblich das Regulierungswerk). The reliability and immutability of a digital contract on a blockchain depend on the one hand on the skills of the developers and programmers. (built-in "back doors" are a "no-go"!) On the other hand, however, those who want to map their business models using a blockchain application are also in demand. This group of people is at least equally responsible for ensuring that the code-based systems they develop comply with the law and other regulations.

To implement the code on a blockchain, a look at existing standards and libraries helps, because it is not about reinventing the wheel, but rather—figuratively speaking—about making a vehicle out of the wheel through new combinations.

5.1   Test Networks and Libraries

There are a variety of cryptographic algorithms and mathematics in blockchain technology. Before transactions can be sent to the blockchain from an application, they must be prepared. Transaction preparation includes, among other things, the definition of accounts and addresses, the addition of required parameters and values to the transaction objects, and the signing with private keys. Applications can be tested using test networks that mimic the "real" service. Errors can be found in this protected environment without affecting the so-called mainnet.

When developing applications, it is better to use verified and tested libraries for transaction preparation than to rewrite code from scratch.

Some of the stable libraries for Bitcoin and Ethereum, with which transactions can be prepared, signed and sent to the blockchain nodes/network, are available as open source.

For Java as an important programming language there are also a wealth of important tools, frameworks and libraries. The following list does not claim to be complete, but only shows a small selection.

- JUnit: The JUnit platform serves as the foundation for launching test frameworks on the Java Virtual Machine. In addition, the TestEngine API is defined for developing a test framework that runs on the platform. The current version JUnit 5 consists of various modules. For more information, see junit.org/junit5/including Usere Guide, Java Doc and GitHub Repository.
- Selenium: Selenium is probably the most popular tool for Java UI testing, allowing you to test your JSP pages without having to launch them in a browser.
  You can test your Web Application UI with JUnit and Selenium. It even allows you to write acceptance tests for web applications. For further information, see seleniumhq. org.
- TestNG: TestNG is a test framework inspired by JUnit and NUnit, but with many new features that make it more powerful and user-friendly, such as annotations, the ability to run your tests in arbitrarily large thread pools with different available policies (all methods in their own thread, one thread per test class, etc.). For more information, see testng.org/doc/.

- Mockito: There are many mocking frameworks for Java classes, such as PowerMock and JMock. Mockito stands out because of its simple API, great documentation, and many examples. Mocking is one of the key techniques of modern unit testing, as it allows you to test your code in isolation and without dependencies. For further reading, please go to side.mockito.org.

The list of test networks and libraries can be almost endlessly extended. Front-end applications also have these corresponding supporting services.

Since this book is not a book for software developers, but for users, only a very rough overview is given of how, for example, the Ethereum blockchain architecture is set up.

If you want to test a little bit yourself how to run a smart contract as a test version over the Ethereum network, then the test environment Rinkeby is recommended (see also note Torsten Horn under DApp).

In addition, you will find some explanations of the Ethereum blockchain below.

5.1.1 Ethereum Blockchain

The Ethereum blockchain is designed similarly to the Bitcoin blockchain. The main difference between the two blockchains is that in addition to the transaction list and the block number, the latest status is also given on the Ethereum blocks. However, the Bitcoin blockchain can "only" send currency units from one sender to one receiver. The Ethereum blockchain allows any type of value exchange through the integration of smart contracts.

A block is basically validated in the Ethereum network as follows (cf. Fig. 5.2):

- It is checked whether the block set in relation to other blocks exists and is valid.
- The timestamp of the blocks is checked.
- The block number and all "roots" are checked for validity.
- The proof of work is checked for validity.
- S [0] is set as the end of the previous block.
- Tx [0], Tx [1], etc. is the transaction list with the number of transactions.
- If a transaction list or an application (Apply) has an error or a time limit is exceeded, an input is considered invalid and rejected.
- The block validation algorithm ends theoretically with S [n], but the miner must be paid who makes his/her hardware available to validate this algorithm. For this reason, before S_Final (executing a transaction), PAY BLOCK REWARD is added.
- It is checked whether all blocks match the origin and thus whether the blocks are valid or not.[9]

---

[9] GitHub Inc. Whitepaper Ethereum, 2017, accessed November 16, 2019.

Fig. 5.2 Validation process on
the Ethereum blockchain. (own
representation based on Fig. on
Ethereum Github)

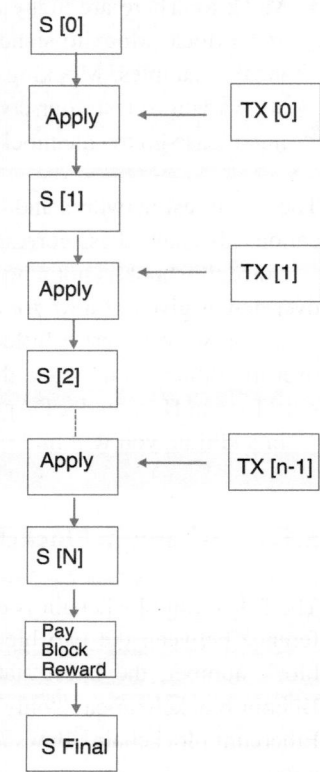

Smart contracts, which are an essential part of the Ethereum blockchain, transport values
that are only released when the conditions of the contract are met.

Smart contracts are usually developed and written in the Solidity programming language, implemented in the Ethereum Virtual Machine (EVM) and executed there. In
addition to Solidity, the following programming languages can be used to develop smart
contracts:

- Geth (GO),
- Go- ethereum (GO),
- Parity (Rust),
- Cpp-ethereum (C++),
- Pyethapp (Python),
- Ethereumjs-lib (Javascript),
- Solidity (syntax).

5.1.2   Geth

Geth is the official client software of the Ethereum Foundation. It is written in the programming language GO. This software includes the following important components:

- Client Deamon,
- Geth Console,
- Mist Browser.

A daemon is a program that runs continuously in the background and provides certain services.[10] The daemon program can forward the specified requirements to other programs or processes if necessary. Every server of websites on the World Wide Web has an HTTPD or Hypertext Transfer Protocol daemon that continuously waits for requests from web clients and their users. If the client (also called node) connects to the daemon within the blockchain, he/she also connects to other clients on the network and downloads a copy of the blockchain. This client communicates continuously with the other clients. In this way, the copies of the blockchain are always kept up to date. The client daemon has the ability to generate blocks, add transactions to the blockchain, validate transactions, and execute them. In addition, it also acts as a server by exposing an API (Application programming Interface) that can be interacted with via Remote Procedure Call (RPC).

The Geth Console is a command line tool that establishes connections between the participating nodes in the network and then allows them to interact with each other. The following actions are, for example, conceivable:

- Create accounts,
- Manage accounts,
- Communicate,
- Confirm transactions,
- Submit transactions to the blockchain,
- Query the blockchain.

The Mist Browser was an integral part of the Ethereum ecosystem for a long time. This was a desktop application that was used to communicate between nodes. All actions that are possible with a console could also be carried out via the graphical user interface of the Mist Browser. Smart contracts could be easily programmed and implemented in many applications via the Mist Browser.

A standard web browser like Chrome, Firefox or Internet Explorer allows access to websites like Amazon, Facebook, Google etc. A similar comparison can be made with

---

[10] Kersken (2019), p. 405.

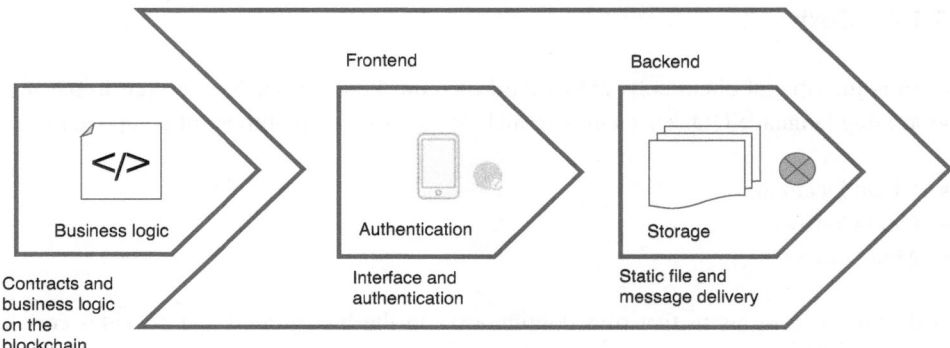

Fig. 5.3 Structure of a DApp. (Figure based on Dhillon et al.)

various mobile apps that are available via Google Play. Similarly, the Mist Browser allowed access to decentralized apps that are available in the Ethereum network.

However, increasing security challenges and the spread of solid alternatives such as Brave, Opera and Coinbase Wallet Mobile have led to the discontinuation of this browser.

5.2 Decentralized Application (DApp)

Applications that are completely decentralized are called "decentralized Applications". The special thing about it is that, for example, a mobile or web app has a back end that runs on centralized dedicated servers; however, a DApp has its back-end code in a decentralized peer-to-peer network.[11] A DApp is a server-less application that, for example, runs on the Ethereum stack and communicates with the end user via an HTML/JavaScript front end that can make calls to the back-end stack (see Fig. 5.3).

DApps have their own set of associated contracts on the blockchain to be used, which are used to encode the business logic and allow for permanent storage.

Its source code is available as open source and can therefore be used freely. Anyone can view the source code and modify it for their own purposes. The source code is decisive for an application: If the source code is changed, the functions of the computer program change.

DApps must store data, reports, and the original source code on a blockchain. This can either be its own blockchain, or one of the DApp platforms developed for this purpose can be used.

---

[11] Dhillon/Metcalf/Hooper (2017), p. 42.

To try out your own idea for a smart contract "simply", you can follow the instructions of Torsten Horn.[12] Each step to be carried out is described on this website, so that even non-programmers have a chance.

5.3 Conclusion

IT and code are always part of the whole. The way in which this increasingly decisive part of our action determines and affects our daily lives will be one of the important research questions of the future.

However, we can observe in our current business and working world that we can be active in shaping it. Traditional and centrally oriented systems that we are used to can be rethought and redesigned. The way decisions are made changes as a result. The network decides, the network evaluates. Therefore, every organization, every company has to deal with the questions of IT with regard to architecture and the principles to be applied. Which processes, internally as well as externally, are linked to the stakeholders on which basis?

Every problem that can be mathematically described is mapped out through algorithms and corresponding programming. With the use of machine learning or deep learning, a program code can learn independently. For this it needs enormous amounts of data to recognize a reliable pattern. In order that there are no undesirable effects in the development of patterns, the data to be used must be reliable and must not create any bias. For example, there are now indications that Google has a misogynistic algorithm that ranks websites of women per se worse.[13] Data capture can thus become a minefield.

It is therefore important to take both the technical (programming) possibilities and the further development including security requirements into account. Developing and testing one's own ideas should be done taking into account governance rules. Then the code and the power it contains can unfold in the sense of "desirable behavior". This can be checked in the corresponding test environment before a new application comes onto the market.

The blockchain technology has set out to overcome centralized approaches. Consensus mechanisms help to define how decision-making takes place in a network and how the execution of the decisions made is to be carried out. A code has a lot of power,

---

[12] Horn (2019): Smart Contracts for the Ethereum Platform. https://www.torsten-horn.de/techdocs/Ethereum-Smart-Contracts.html#Installation-Geth-Solc-Test-Blockchain, accessed on October 15, 2019.

[13] In summer 2019, this disadvantage was reported several times, for example on Golem.de, which offers IT news for professionals.

but every code, no matter how powerful, needs an environment in which it can act. How this environment is designed, central or decentralized, federal or monarchical, is up to us as decision-makers.

References

Dhillon V, Metcalf D, Hooper M (2017) Blockchain enabled application, understand the block-chain ecosystem and how to make it work for you. Apress/Springer Science, New York
de Filippi P, Wright A (2018) Blockchain and the law: the rule of code. Harvard University Press, London
GitHub Inc (2017) Whitepaper Etheureum. https://github.com/ethereum/wiki/wiki/White-Paper. Zugegriffen: 16. Nov. 2019
Horn T (2019) Smart Contracts für die Ethereum Blockchain. https://www.torsten-horn.de/tech-docs/Ethereum-Smart-Contracts.html#Installation-Geth-Solc-Test-Blockchain. Zugegriffen: 15. Okt. 2019
Kersken S (2019) IT-Handbuch für Fachinformatiker, Der Ausbildungsbegleiter, 9., erw Aufl. Rheinwerk, Bonn
Lessig L (2006) The code version 2.0, 2. Aufl. Basic Books, New York
Prewitt M, McKnie S (2018) Blockchain communities and their emergent governance. https://medium.com/blockchannel/blockchain-communities-and-their-emergent-governance-fdf24329551f. Accessed: 12. Oct. 2019
Sauerwein F, Just N, Latzer M (2017) Algorithmische Selektion im Internet: Risiken und Governance automatisierter Auswahlprozesse. kommunikation@gesellschaft 18:1–22. https://nbn-resolving.org/urn:nbn:de:0168-ssoar-51466-4
Weill PD (2004) Don't just lead, govern: how top-performing firms govern IT. MIS Quarterly Executive 3(1):1–17

The Next Hype?

Abstract

Disruptive! New! Better! Adjectives that are mentioned in connection with blockchain technology are endless. They often describe a hype, and blockchain technology has seen some hype in its short existence. The different token models that can be used for different purposes are described below. The initial offering models are presented that are used in addition to the issuance of the respective tokens for capital allocation. Since new variations of token models are added over time, a key basis, namely the so-called ERC20 token, is considered in more detail in connection with the ERC721 token of Ethereum. This makes it possible to better understand the new trend 2021: so-called non-fungible tokens (NFT). Furthermore, it is explained how digital central money (CBDC; Central Bank Digital Currency) is to be located in this context.

A preliminary highlight was the year 2017. Not a day went by without the media reporting on Bitcoin, blockchain and the so-called initial coin offering (ICO in analogy to the IPO [initial public offering] for stock corporations). The price of a Bitcoin rose to over 20,000 USD at the end of the year, only to fall sharply in the course of January 2018 (at the end of January, the value of a Bitcoin was "only" 9628 USD). On the one hand, so-called follow-on effects were mentioned as the cause of the price decline. On the other hand, regulatory announcements have led to the dampening of the existing gold rush atmosphere. For example, South Korea has introduced new rules for cryptocurrency trading, and the Security and Exchange Commission (SEC) decided in summer 2017 that ICOs are to be assigned to the securities business. This regulatory development dampened the hype around the initial coin offering.

However, this hype (and others) was driven by investors who are willing to participate in business ideas at a very early stage. Examples of this new wave of innovative business

K. Adam, *Blockchain Technology for Business Processes*,
https://doi.org/10.1007/978-3-662-65818-5_6

ideas in a new innovative environment were, for example, Mark Zuckerberg, Elon Musk, Jack Ma, all of whom believe in their vision and have developed it into large companies. Every company that offers a token model via a blockchain promises very early participation in the perhaps next so-called unicorn.[1]

The basic idea behind tokenization is to give small investors the opportunity to participate in innovative and still very early business ideas.[2] Usually, this early investment phase is accompanied by professional engagement—venture capital/risk capital. These are risk capital companies that collect money from their investors to invest in young, mostly technology-oriented companies. In return for the provision of capital, the risk capital companies demand a very high return taking into account an so-called exit strategy. In the classical financial world, the exit for these companies means selling their shares after the promoted company has been brought to the stock exchange.

Small investors usually do not have access to such investment opportunities because the tranches are larger than the amount a small investor can pay. The bundling of small sums to participate in larger investment projects is also often referred to as crowdfunding. However, this crowdfunding is rather used for local projects. Coins and tokens, on the other hand, can be used worldwide and virtually borderless.

The acronym TGE is often used in connection with ICO. The acronym stands for Token Generating Event and can be used as synonymous with Initial Coin Offering as well as the term Token Sale. They all have in common the issuance of tokens to finance a business idea. The token represents a value corresponding to the business idea. This can be the value of the created own currency or a certain right.

Background Information

Two terms are used in connection with this hype: "coin" and "token". For better differentiation, the terms can be defined as follows (see Table 6.1).

A "coin" represents an explicit blockchain with its defined technical properties. So a Bitcoin belongs to the Bitcoin blockchain, a Neo to the Neo blockchain, etc.

A "token" is not a representative of a blockchain, but rather a token uses the technical properties of a blockchain.[3]

Applied to the ICO hype of 2017, this means that instead of shares, tokens are now created which can be acquired within the ICO. These tokens can represent different things (see sect. 6.4), but in the end they represent a company's own cryptocurrency. Overall, this terminology represents an intended approach to the initial issuance of shares (i.e. IPO), but no shares in the company are sold in the token sale. The investor rather invests

---

[1] In business, a unicorn is a start-up with a market value of over one billion US dollars before the actual classical stock market listing. In 2019, the Chinese company Toutiao (Bytedance) is in first place with a valuation of 75 billion US dollars, followed by the American company Uber in second place with 72 billion US dollars.

[2] Gründerszene

[3] Bogensperger et al. (2018), p. 73.

Table 6.1 Gegenüberstellung coin token

Coin	Token
Own cryptocurrency	Depending on the technology of a cryptocurrency
Representative of larger projects	Use in the Initial Coin Offering as a digital voucher
More expensive to program, higher level of difficulty	Easier creation through use of existing technologies (e.g. Ethereum)
A coin only has the function of a means of payment and represents a unit of currency	Can represent different functions
Only needs its own platform	

in a business idea and allows the start-up to develop this business idea into a product with the financial means he/she provides. In return, the investor receives the token, which can represent different properties (see sect. 6.4). Overall, Initial Coin Offerings have become one of the most popular financing methods for start-ups in 2017.[4]

The project-based corporate financing of an ICO is now taking on new forms of expression.

6.1   Initial Exchange Offering (IEO)

In 2017 and 2018, a large number of successful ICOs came onto the market, but also an incredibly large number that were unable to meet their promised returns. Therefore, ICOs have fallen into disfavor due to underperformance, lack of investor interest, and regulatory uncertainty. In principle, it is a good approach to raise funds from investors across borders. Therefore, this ICO approach is being modified, and in 2019 it is now more common to hear about IEOs.

While the IEO trend began as an Asian phenomenon, led by large Asian exchanges, the model is now being adopted by large, reputable US exchanges. Bitfinex has recently launched the Tokinex IEO platform, and Coinbase is also interested in launching an IEO platform as well as a STO emissions product (Security Token Offer).

While Initial Coin Offerings are a form of peer-to-peer fundraising typically conducted on a project's own website or with the support of a specialized platform, Initial Exchange Offerings are token sales hosted by a third-party exchange.

For the retail investor, an IEO has several risk-reducing advantages over an ICO: The exchange conducts due diligence to ensure quality, credibility, and potential of the

---

[4] Hahn/Wons (2018), p. 3.

project. There is an immediate secondary market at the time of issuance to guarantee
liquidity. The exchange has a vested interest in only offering sustainable investment
opportunities due to its own reputation risk.

Like ICOs, IEOs are a form of early-stage investment available to retail investors.
Instead of purchasing equity in a company, retail investors use IEOs to invest in the
potential benefits of a company. In return, they receive a token that represents the benefits promise. Companies that already have a product are therefore at an advantage.

Blockchain startups with first products can use this new form of financing in the seed
phase as a complement to traditional venture capital.

The research platform BraveNewCoin predicts an increasing number of IEOs for the
year 2020. The low interest rate environment makes it difficult to find yield, and investors' interest in cryptocurrencies as an alternative investment is growing.

6.2   Initial Futures Offering (IFO)

In September 2019 the so-called Initial Futures Offering (IFO) came into the picture,
which stems from the findings of the ICO and IEO waves.

As the name Futures already says, these are tokens that do not even exist at the time
of purchase. These are derivative financial contracts that oblige the parties to trade an
asset at a fixed future time and price.

The argument for an IFO is that more diversification is possible because a greater
number of instruments are available.

In early autumn 2019, the regulatory requirements are less strict in comparison to an
ICO or even IEO.

The leverage effect present in futures offers the potential to increase profits by a multiple. The reason for this is that the guarantee coverage is often much lower than the
underlying asset. However, the leverage effect also applies to potential losses, which
could then be many times higher than expected.[5]

It is expected that futures will provide a more realistic price discovery for a token.
The view of the business idea, progress and success is reflected in a future as an expectation value.

If "normal" futures are already highly risky and therefore more suitable for professional investors, I consider this digital investment opportunity to be at least as risky. It
is the future bet on a digital (and thus non-haptic) and so far non-existent product. Just
as bets on future harvests have a high risk potential, the risk of suffering a total loss on a
blockchain-based future token is high.

---

[5] Becker (2012), p. 11.

6.3   Token Economy and New Business Areas

As the understanding of what a blockchain can do in addition to transferring currency units grows, so does the development of the token economy.[6] The first push is inextricably linked to the success of Ethereum. Ethereum was the first platform that allowed hundreds of smart, contract-based decentralized apps (DApps) to be run and the associated token to be issued. By way of comparison: just as Microsoft Windows was for the computer, DApps are for the blockchain world.[7]

In this blockchain world, new organizations are emerging that set their own conditions and rules and thus create new, self-sustaining mini-economic systems. A provider of a token in this context can freely decide which rights and obligations are to be associated with a token issued by him. Tokens need not contain membership, information, control or voting rights.

It is already apparent today that the token usage relationships are much more important than the design of the underlying crypto-economy. There is no perfect token. The business model must be viable—otherwise the best token is of no use.

However, this so-called token economy is already influencing the linking of the real with the digital world. There is hardly any real value that cannot be digitized and thus transferred into a token model. And just as unlimited as digital values can be created from real and real values, so are the possible uses and the corresponding business areas.

In addition to the possibility for young companies to raise significant financial resources through the issuance of tokens, based on their own business idea, this new economy requires new activities. Networks—computer networks, developer platforms, marketplaces, social networks, etc.—have always been a strong part of the Internet's promise. To be successful as a network, it requires a critical mass. This can be achieved more easily with token models, as tokens can be designed in many ways (see also Sect. 6.3).

In addition, token models can create incentives for network participants by aligning network participants to work together towards a common goal—this can be, for example, the growth of the network and or the appreciation of the token. This orientation is, by the way, one of the reasons why Bitcoin continues to grow and thrive despite the skeptics.

Background Information
The design of the token depends on the business purpose behind it. The content and conceptual design as well as the target of how much money should be raised with the issuance of the tokens is to be thought through in advance.

Creating your own tokens is relatively easy, as you can use the Ethereum standard. For example, the ERC20 standard from Ethereum is mentioned here.[8]

However, good planning and setting the right goals are important.

---

[6] Voshmgir (2019), p. 3.

[7] Casey/Vigna (2018), p. 99.

[8] ERC stands for "Ethereum Request for Comment", and the number 20 is used as an identification number to distinguish this standard from others.

The sheer number of tokens is critically important to consider. The website Coin Market Cap lists almost 9950 different currencies and, correspondingly, almost 9950 different coins as of Juli 2022.[9]

An—also internationally valid—identification and classification is still pending. The consequence is that on a national level, the assessment and technical classification vary greatly. The first step in this direction is taken by the Principality of Liechtenstein, which passed the world's first blockchain law in October 2019, which came into force on 1 January 2020. This creates the legal certainty that protects investors and consumers better and at the same time ensures appropriate monitoring of the various service providers in this environment. It remains to be seen to what extent this contributes to global standardization.

The most common types of tokens are described below.

6.4   Token Typology

There are different classes of tokens, and they can roughly be divided into security tokens (security tokens) and tokens with a functionality promise (utility tokens). The determination of the rights has an immediate effect on the legal classification of the token. A token must now also comply with supervisory requirements.

The distinction is important in terms of the functionality of the two variants. Above all, the financial and stock exchange supervisory authorities attach importance to the fact that they are declared correctly. They were the ones who made this division necessary.

The Securities and Exchange Commission (SEC), founded in 1929 during the great stock market crash, has become aware of this development with the increasing number of initial coin offerings (ICOs). The SEC's main concern is to protect investors and ensure that the trading of securities is legally safe and controlled. The SEC intervened in this market when it became clear that a large number of offers did not have a viable business model and that some of them were fraudulent.

Securities are tradable financial assets such as bonds, options, stocks and warrants. Security tokens meet the same requirements and are therefore considered securities. For example, both the SEC and BaFin have several tests to determine into which category a token falls. The best known is probably the Howey test, which summarises the requirements.

A token is a security token if the following points are all fulfilled:

- The token is an investment.
- The investment goes to a company or group of companies.

---

[9] Coin Market Cap (2019). https://coinmarketcap.com, Accessed: 24.July 2022 (in 2019, when this book first time was published in German this platform "only" provided information about almost 3.000 coins. This means within three years the number of coins has tripled.)

Token types

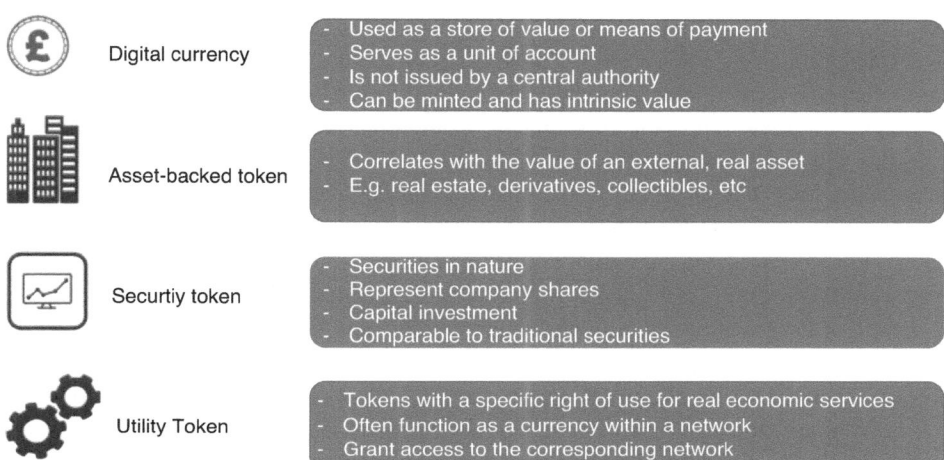

Fig. 6.1 Token types

- The investor expects to make a profit by acquiring the token.
- The expected profit is generated by the work of third parties.

If a token meets these 4 points, the company must meet all the requirements that apply to traditional securities. These include the prospectus requirement, ad-hoc reporting requirement and liability for incorrect information. If a company does not meet these obligations, significant penalties may result.

Many companies see this form of token offering as a sensible alternative to other means of raising money (cf. Fig. 6.1; see also Sect. 7.6).

The token types listed in the figure are explained below.

6.4.1 Payment Token or Digital Currency

The term "digital currency" may be misleading, because (currently) these are not currencies recognized by central banks. Nevertheless, these currencies also have the classic features of currencies, albeit to a lesser extent:[10]

Exchange function: An exchange function is given when an owner of a good is willing to exchange this good for the corresponding currency. In terms of traditional

---

[10]Altmann (2003), pp. 80 ff.

currencies (so-called fiat money)[11] we experience this exchange function constantly. We buy our bread from the baker and put the corresponding amount of money on the counter. Similarly, this is the case with a digital currency unit. This is usually obtained by exchanging one's fiat currency for digital currencies. The considerations of creating money from the mining process or receiving a digital currency token through other incentive programs represent an exchange: computing power for Bitcoin or other services for a payment token.

Accounting unit: The presence of this function makes it possible to assess the goods and services to be exchanged. In this way, heterogeneous goods can be made comparable. For example, you can buy journalistic content, such as newspaper articles, on the Satoshipay website via so-called micro-payments with a fraction of a Bitcoin. The length of the paragraph determines the price to be paid.[12]

Preservation of value function: In economic terms, a currency must preserve its value over a certain period of time, in other words, it must be conserved. Digital currencies, on the other hand, are very volatile, i.e. they fluctuate greatly in value. For this reason, the preservation of value function is denied to digital currencies; they may represent a value, but they are not able to maintain this value stable over a longer period of time.

Even though digital currencies are not recognized as currencies by central banks in the Western world such as the European Central Bank, the Bank of England or the Federal Reserve, there is still a lot of movement in this market. With Facebook's currency "Libra"later known as "Diem", it becomes clear that digital currencies have the potential to become "system-relevant" and thus to pose completely new challenges to the money market as we know it. The Libra Foundation based in Switzerland explains in the white paper belonging to the currency that exchange rate fluctuations (and thus a high degree of volatility as with the other digital currencies) can be greatly reduced, as Libra is backed by a "basket" of bank bonds and short-term government bonds in different, official hard currencies.[13] This suggests the same security and confidence in a currency issued by a private company as is otherwise claimed only by central banks. If Libra were used as a daily means of payment, it could be much more than just another payment service provider that includes people currently excluded from the financial system. Libra could create its own economic cycle that would make central banks obsolete worldwide. At least the Libra white paper leaves this possibility open.

Politicians and regulatory authorities worldwide have countless questions and just as many reservations about this proposal. Because no state wants to see its right to issue currency restricted by a private company, many governments are investigating state digital currencies: for example, the Swedish central bank is researching "E-Krona",[14]

---

[11] Fiat (Latin) can be translated as "let it be"; this is central bank money.

[12] Interview with Meinhard Benn on Gründerszene.

[13] The Libra Blockchain (2019), White Paper, pp. 23 ff.

[14] See Riksbank (2019): https://www.riksbank.se/en-gb/payments%2D%2Dcash/e-krona/.

Federal Minister of Finance Olaf Scholz is thinking aloud about the e-Euro, and it is said that the Chinese central bank wants to introduce the digital yuan[15] in 2020. Annotation: In January 2022, the cryptocurrency Diem backed by Meta Platform Inc. officially announces its end.

6.4.2   Asset-Backed Token

This token is subject to a real asset, and it can be understood as the digital version of asset-backed securities (ABS).

Asset-backed securities are securities that derive their income from one (or more) underlying asset (s). These can be real estate, receivables, other assets such as paintings or financial securities.[16] Similar to a fund, the buyer acquires a part of the total asset value. The term "fund" is described in Germany in § 1 Investment Code (KAGB). According to § 1 para. 1 sentence 1 KAGB, "investment assets [...] any organization for common investments that collects capital from a number of investors in order to invest it in accordance with a specified investment strategy and which is not an operational company outside the financial sector."

Asset-backed tokens adopt this approach and allow investors to participate with small amounts in an asset. The special feature of this digital participation is that fractional tokenized assets make a whole range of new asset classes possible and that these assets become more fungible and liquid.[17]

The asset-backed token represents a claim that is linked to the corresponding asset and its income.

6.4.3   Utility Token

The utility token is a functional part of a product or service of a start-up. Initially, the utility token was the most commonly used token form in an ICO. This type of token allows the owner to access a specific service that is defined and set in the smart contract in advance and to pay for the retrieved service with the token. Sometimes the analogy is made to tokens in a casino or poker game: In both cases you have to buy special game chips to participate in the actual game.

---

[15] See Chinadaily (2019): http://www.chinadaily.com.cn/a/201909/04/WS5d6eefa1a310cf3e3556985a.html; accessed September 11, 2019.

[16] Cf. Berk/DeMarzo (2019), p. 735.

[17] Cf. Voshmgir (2019), p. 24.

Utility tokens do not represent a security-like asset, unlike security tokens. They are also not a classic cryptocurrency. They are "only" the promise of a product or service that is often still to be created.

Overall, they are not subject to as strict regulations as a security token.

The strength of the utility tokens lies in the binding to the users as participants in a network. The community of this start-up can be built up via the functionality of the utility token. As soon as the critical mass mentioned above is reached, the token gains intrinsic value and becomes attractive for participants who were not represented in the network before.

Supporters of utility tokens are convinced that in the future networks and platforms will be more democratic with the help of these tokens. Today's well-known platform providers such as Amazon or Alibaba have become such large market participants because they are centrally managed. Blockchain networks with utility tokens can define mutual incentives in a decentralized manner. Transparency is considered one of the success components.

Start-ups that use utility tokens imitate the existing processes of companies or institutions to a certain extent. However, the start-ups set their own rules. The application of these rules is then integrated into the system via a smart contract. So an investor invests with a utility token in the vision of the corresponding company and in the expectation of future market developments.

6.4.4   Security Token

A security token behaves like a security and is also assessed as such by the German Federal Financial Supervisory Authority (BaFin).[18] This can mean that the token is linked to a profit and sales promise. If repayment obligations can be related to the token, this is also referred to as a security token, which is constructed similarly to a security. The same applies if the token holder has equity rights in the assets of the corresponding start-up or if the token reflects a money investment. Whether the token is identified as a security token depends on the Howey test described above. If a token does not pass this test, it is a strong indication that it is not a security token that must comply with traditional securities regulations with all the disclosure and registration requirements.

However, if it is a security token, this implies more rights for the token holders; all security-like rights can be secured via a blockchain.

A securities prospectus becomes due at the latest when the security token is offered to private investors. This securities prospectus must be approved by BaFin. The small, so-called simplified version in the form of a securities information sheet, which should only be three DIN A4 pages in size, allows for a maximum capital allocation of 8 million

---

[18] BaFin, 2018, BaFin Perspectives.

euros. These security tokens can then only be traded in Germany—while security tokens with a securities prospectus can be traded throughout Europe. It should be noted that security tokens are equipped with the properties of securities and are therefore, like a security, a fungible (exchangeable and tradeable) financial instrument that represents actual money value. Because security tokens are usually backed by real assets, the reputation of the issuers of such tokens grows. This security reduces the investment risk of a total loss for the investor.

Other advantages are:

- The acceleration of processes, since investors and token providers can exchange directly with each other without an intermediary.
- The avoidance of intermediaries reduces costs, and it is expected that the widespread approach of smart contracts will further simplify and streamline future trade.
- So-called "fractional ownership", because a high-quality asset can be divided into smaller units than is currently common. This in turn makes wider participation possible—also from small investors.
- Wider investment spectrum: Companies can reach a wider investor base by offering products and services on the Internet.
- Wider participation: An STO is not limited to a region or country, but can rather operate global trade over the Internet.

Disadvantages are also to be named:

- Restriction of the secondary market: Security tokens are usually not freely transferable. There is often a so-called "whitelist" that indicates who can potentially be an authorized token holder or is already accredited and can therefore buy the token. However, this can severely restrict trade for the investor.
- Accreditation: Many STOs require investors to be accredited in order to participate. This includes that an investor must go through the "know your customer" process as well as the one for the money laundering law. In principle, this is to be welcomed so that this form of investment is not associated with money laundering and terrorist financing. But even if much of this can be automated, the effort to be made by the investor could be seen as hindering and therefore discouraging.

 The security token market is still very young. Precedents that can be referred to are available, but these cases are—like the market—very young; how these cases will fare under difficult conditions is not yet certain. The legal design of a token should therefore always be done in consultation with the Federal Financial Supervisory Authority (BaFin). However, it is also important to realize that in Germany BaFin does not carry out an substantive review of whether a business model makes sense or not. BaFin only checks whether a token issuer has complied with the extensive legal requirements.

Fungible token   Non-fungible token

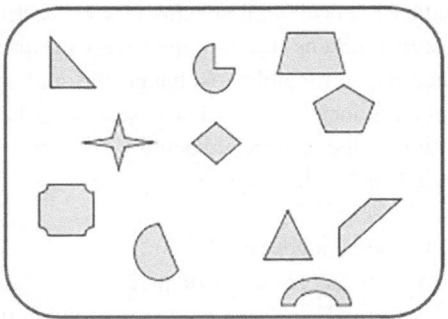

Each token represents the same value   Each token is unique and one-of-a-kind
and is therefore exchangeable and
replaceable by another token.

Fig. 6.2 Difference between fungible and non-fungible-token

Security Token Offering (STO), that is offering a security token for purchase by investors, follows the same motivation as an ICO. Funds are raised for corporate financing.

6.4.5 Non-Fungible-Token (NFT)

In 2021, the general public was introduced to the term NFT. The decisive event was the first-ever auction of a digital collage by artist Mike Winkelmann, known as Beeple, at Christies auction house.[19] This monumental work of art, which dates back to May 2007, was auctioned off in March 2021 for over $69 million—a sum that is usually only achieved at auctions for works by Jeff Koons and David Hockney.

So what are NFTs and how can they be used?

NFT stands for Non-Fungible-Token and allows assets to be represented on a blockchain and thus ownership to be assigned (Fig. 6.2).

In contrast to the fungible tokens that have dominated so far, a non-fungible token can be assigned unique and identifiable characteristics. This makes a non-fungible token unique and not interchangeable. These tokens are defined, for example, by the ERC721 (Ethereum Request for Comments 721) standard. More information about the token standards created by Ethereum can be found in Sect. 6.4.

---

[19] https://www.christies.com/features/Monumental-collage-by-Beeple-is-first-purely-digital-artwork-NFT-to-come-to-auction-11510-7.aspx, accessed 10.09.2021; 10:45.

The value of an NFT is based on the asset that the NFT represents. This means that every NFT has a unique identity that distinguishes it from others. This identity and uniqueness make it possible to create a demand surplus on a market that is reflected in significantly higher prices per token.

A beautiful analogy makes the idea of NFTs clear: assigning ownership can be compared to the artist's signature under a picture. An NFT is therefore the digital, modern version of this method. The potential is rated very high: from fake news to the unauthorized use of faces and pictures.[20] up to pieces of music—everything can be represented by a Non-Fungible Token.[21] And this is just one aspect if you look at the artist scene.

Smart contracts based on the ERC721 standard serve as the technical basis, which enable the creation of NFTs (see sect. 6.5).[22] This ERC721 standard enables the implementation of common basic functions for the management of assets. It is to be understood as an extension of the ERC20 standard, which was already developed at the end of 2015, in order to enable fundraising during the actual development phase via the integration of smart contracts.[23]

The following will explain these two standards in order to describe the potential.

6.5   ERC20 in Connection with ERC721

At first glance, when you see this chapter heading, you get the feeling that everything here is very cryptic and needs to be deciphered first.

---

[20] Generative Adversarial Networks (GAN) allows data to be generated from two competing Artificial Neural Networks (ANN), one of which generates data that looks real and is classified as real or artificial by the second ANN. The learning curve of these formats is steep and the artificial data becomes more and more like the real one. As a result, the human eye can hardly perceive the differences and it is therefore no longer possible to distinguish whether the photo or video is real or generated by a computer algorithm. Interestingly, we users of apps like Snapchat or Whatsapp have contributed to the fact that an infinite number of data (= photos [and also portrait photos]) are available. By using so-called face filters, we have given our consent to use our images and the companies have "fed" their algorithms with our images. As a result, texts, videos and photos created in this way are no longer distinguishable for human eyes and ears—and thus also allow the space for unethical uses of these created data. In February 2019, therefore, software engineer Phillip Wang set up a website to raise awareness of the power of machine learning and the potential risks involved. The website entitled "This Person does not exist (https://thispersondoesnotexist. com)" published by GANs artificial portraits of human faces created.

[21] Economist Newspaper Limited (London), March 18th 2021, Online edition: https://www.economist.com/finance-and-economics/2021/03/18/non-fungible-tokens-are-useful-innovative-and-frothy, accessed on 08.09.2021; 19:45.

[22] Zheng/Dai/Wu (2021) Blockchain Intelligence, p. 17 with p. 19.

[23] Vogelsteller/Buterin (2015): https://eips.ethereum.org/EIPS/eip-20, accessed on 14.09.2021; 15:30.

Section 6.3 explains that ERC stands for **E**thereum **R**equest for **C**omments and the corresponding number refers to a corresponding standard. Many tokens are generated through this mechanism. The following explains the two standards currently used for the creation of tokens, including their applications.

The ERC20 standard is based on a corresponding smart contract when creating tokens, through which the number of tokens is determined. In addition, this standard allows with the corresponding smart contract to keep track of who owns a certain token at any time. [24] This standard, created in 2015 with simple smart contract instructions, has since been the basis for many tokens traded in the market.

The following features are specified with the standard:

- Token Name
- Token Symbol
- Decimal number (up to 18 digits)
- Total supply
- Balanceof
- Transfer
- Transfer (from)
- Approval
- Allowance[25]

With these features, tokens can be individually customized to an almost infinite extent.[26] To create such an ERC20 token, no separate blockchain community is required to enable the transactions. Rather, the tokens including their rights are traded via the Ethereum blockchain—quasi on top.[27] Typically, ERC20 tokens are fungible, i.e. interchangeable with other tokens of the same type and quality and each token can be fractionalized.[28]

This is different with the ERC721 token: this token represents a unique asset and is therefore not interchangeable with a token of the same type and quality. This extended possibility of the ERC20 standard enables NFTs (Non-Fungible-Token). The first popular application are the CryptoKitties.[29] In this game, the user breeds virtual cats that he/she can sell. Each of these virtual cats is unique and represents an asset that cannot be shared with anyone.[30]

---

[24] De Filippi/Wright (2018), p. 249.

[25] https://wirexapp.com/blog/post/erc20-vs-erc721-whats-the-difference-0341, accessed on 15.09.2021; 21:30.

[26] Hellwig/Karlic/Huchzermeier (2020: p. 42).

[27] Casey/Vigna (2018), p. 102.

[28] https://ethereum.org/en/developers/docs/standards/tokens/erc-20/, accessed on 18.09.2021; 10:20 PM.

[29] https://www.cryptokitties.co, accessed on 18.09.2021; 10:45 PM.

[30] https://www.deutschlandfunkkultur.de/hype-um-cryptokitties-die-gamifizierung-der-block-chain.1264.de.html?dram:article_id=424579, accessed on 18.09.2021; 10:30 PM.

The ERC721 standard, based on the ERC20 standard, allows the implementation of non-fungible tokens and has the following properties:

- Non-divisible
- Non-fungible
- Additional possibilities for individual design of the token
- Extended contract design possible, i.e. a contract can contain one or more tokens
- Traceability of the owner history back to the origin.

These properties currently stimulate the imagination of many and the hype around NFTs seems to be developing beyond the art market. So it remains exciting.

A third ERC standard provided by Ethereum enables functions that were not included in the other two standards, such as emergency recovery in the event of a loss of the private key.[31]

6.6   Digital Central Bank Money (Central Bank Digital Currency; CBDC)

Since the publication of the white paper "Bitcoin: A Peer-to-Peer Electronic Cash System" by Satoshi Nakamoto[32] attention has been constantly growing. While initially the technology and the privately oriented financial sector came up with new "currencies" and applications, it was at the latest since the announcement by Facebook,[33] to create its own currency, it is clear that central banks have to deal with the topic. This also applies to the European Central Bank (ECB), which, as guardian of the currency, is responsible for maintaining financial stability in the euro area. Money and payment transactions are fundamental functions of a society and the ECB is independent of instructions.

Monetary affairs have been the driving force of trade and growth for centuries. With the increase in global financial transactions and the need to transform these transactions safely into the digital world, the use of new technologies is becoming interesting. Simply put, processes along the supply chain could be covered by a CBDC in the same way as a European consumer's purchase at the supermarket checkout or online shopping. This requires that industry, trade and users have access to this digital currency in the same "simple" way as they currently have access to cash. Digital central bank money is to be a central bank-backed payment instrument and therefore a legal tender.[34]

---

[31] https://ethereum.org/en/developers/docs/standards/tokens/, accessed on 18.09.2021; 23:00.

[32] https://bitcoin.org/bitcoin.pdf, viewed on 16.09.2021; 11:30.

[33] https://www.br.de/nachrichten/netzwelt/was-wurde-aus-facebooks-kryptowaehrung,ScYrHUZ, viewed on 16.09.2021; 11:40.

[34] https://jamaica.loopnews.com/index.php/content/global-interest-proposed-introduction-cbdc, accessed 16.09.2021; 12:15.

In summer 2020, the European Central Bank decided[35] to dedicate itself to the "exploration" of the topic, because a simple transfer of the euro into a digital variant is not enough in view of the complexity of the economies involved. In addition, there are different requirements for the national payment system within the European countries.[36] However, ignoring it is not an alternative, especially since central banks around the world are working on solutions (e.g. since 2014 China and the e-Yuan,[37] Sweden and the e-krona,[38] Bahamas and the Sand Dollar[39] etc.).

For companies, the question arises of how these considerations fit into their own profile. For example, will companies be expected in the future to enable customers (regardless of whether they are industrial or retail customers) to pay with digital currencies? If it is not a payment instrument guaranteed by a central bank, how will, for example, the volatility of the digital currency affect things?[40] Therefore, it is also important for companies to deal with and understand digital currencies.

However, there are differences that need to be considered. Banks, in comparison to the crypto community, speak of accounts as bank accounts. If the holder of a bank account wants to transfer an amount from A to B, he/she must first identify himself/herself as the owner of this account (and owner of the amount lying on the bank account). This identification can be carried out, for example, by means of a PIN and this is considered as proof in order to transfer the money on the account. With a token this is not necessarily necessary. With a token one does not have to prove that one owns the account, it is only necessary to prove that the token is real.

It is similar with cash: everyone who receives cash is entitled to assume that this cash is real. It is often irrelevant to a cash recipient which person hands over the cash to him/her. The only important thing is that the 20 €, which are used for payment, are real and that the recipient of this money can also spend it again (as a reminder: counterfeit money is confiscated immediately upon discovery. If someone knowingly brings counterfeit money into the economy, this is a criminal offense under German law that can be punished with up to five years in prison[41]).

---

[35] European Central Bank (2020): Report on a digital euro.

[36] https://bankenverband.de/themen/europas-antwort-libra/, accessed 16.09.2021; 19:30.

[37] https://www.db.com/news/detail/20210714-digital-yuan-what-is-it-and-how-does-it-work?language_id=1, accessed on 16.09.2021; 19:55.

[38] https://www.riksbank.se/en-gb/payments%2D%2Dcash/e-krona/, accessed on 16.09.2021; 20:00.

[39] https://www.sanddollar.bs, accessed on 16.09.2021; 20:05.

[40] Example: A supplier in industry supplies an OEM. This agrees to pay part of the invoice in Bitcoin. If the Bitcoin rate fluctuates sharply during the course of a day, this can lead to unexpected losses or additional profits. These fluctuations make it more difficult to accept, and an e-Euro issued by the central bank could guarantee missing stability.

[41] $ 147 StGB.

Cash as physical cash can be understood and defined as so-called bearer instrument.[42] And Cash is the only form that is generated by the European Central Bank because it has the monopoly on banknotes.[43]

Giral money, which is issued by commercial banks, is not a legal tender. What does that mean? Giral money is money that the bank then issues when the bank grants a loan, or the bank buys an asset (e.g. bonds). Giral money is "only" a liability and/or a payment promise of the bank.[44]

These relationships have worked well so far. The ongoing transformation of digitization and the global development of cryptocurrencies are calling for an expanded mandate for central banks, including the ECB. According to the ECB's announcements, there will be a coexistence of currencies—the user can decide with which payment instrument he/ she wants to pay. This sounds simple at first, but here too the problem lies in the details. The top priority will be to design and distribute a digital euro, without prejudging the decision to actually introduce it. In a two-year investigation phase, the ECB will analyze how to ensure that a digital euro is available as a risk-free and efficient unit of digital central bank money, with the help of so-called focus groups and expert teams.[45]

The attention the European Central Bank is devoting to the topic should also serve as a reminder to companies that changes are coming in payment traffic. The multi-layered nature of the topic—from the development of its own tokens to privately generated cryptocurrencies to the question of whether there will be a digital equivalent to the euro (or even a programmable, digital euro that can incorporate smart contract functions)— should not discourage, but rather inspire: how can small and medium-sized enterprises in particular use these developments to (re)position themselves in competition.

6.7   Dezentralized Finance (DeFi)

Since 2020, the term "Defi" (Dezentalized Finance) has evoked enthusiasm or strong criticism. DeFi stands for various financial services in a decentralized ecosystem that controls itself via decentralized smart contract applications (DApps) and has no central

---

[42] https://wirtschaftslexikon.gabler.de/definition/e-geld-34967, accessed on 21.09.2021; 13:15.

[43] https://www.bundesbank.de/de/service/schule-und-bildung/erklaerfilme/wie-entsteht-geld-teil-i-bargeld-613640, accessed on 21.09.2021; 13.20.

[44] For example, on your bank account there are 1000 €. As the account holder, you now have the right to exchange this number 1000, which is stated on your bank statement, at the bank for 1000 € cash. So you could say that you have a voucher for payment, which you can of course redeem in case of doubt.

[45] https://www.ecb.europa.eu/press/pr/date/2021/html/ecb.pr210714~d99198ea23.de.html, accessed on 21.09.2021; 13:40.

authority. With DeFi, intermediaries of any kind are to be avoided because the system is open to all participants in the same way. And the hype is huge, which also relieves criticism. The criticism is based on the concern that few sustainable business models will flood the market and eventually a lot of money will be burned.

A closer look is still worth it to understand this trend. First of all, some numbers reveal the great interest:

> The market capitalization is approximately 83.4 billion USD in September 2021,[46] which represents an increase of more than 400%.[47] The magazine *Economist* explains in its issue of 18 September 2021 that the value of transactions that have been processed through the Ethereum Blockchain is approximately 2.5 billion USD in the second quarter of 2021. In order to be able to assess this amount, the Economist refers to Visa with a similar value in the second quarter of 2021.[48] Defi providers promise that the supply and demand for financial resources can be orchestrated through smart contracts and protocols. All of this happens—as always in the blockchain area—peer-to-peer, that is (actually) without intermediaries. In 2021, five areas are crystallizing in which a lot of development can be observed:

- **DEXes**: this acronym stands for "decentralized Exchanges" and these are the platforms on which the investor can exchange his/her fiat money (e.g. Euros, Dollars) for cryptocurrencies. These decentralized exchanges are supposed to enable trading directly between buyers and sellers via their respective wallets, without the need for an intermediary such as a broker.
- **Payment**: The providers of Payment-Solution are striving to solve the traditional problems in the current payment systems because these solutions should enable payments to be processed faster and more cheaply.
- **Lending/Borrowing**: Owners of cryptocurrencies can earn attractive interest payments by lending their tokens.
- **Assets**: Through the tokenization of assets, it is possible for an owner to manage and manage his/her assets in a previously unknown way. In addition, this approach allows the owner to make his/her assets available to others as a digital investment, whether as security, investment or otherwise.
- **Derivatives**[49]: Numerous derivatives can be created and secured via smart contracts on the Ethereum blockchain without the need for an intermediary. For example, this approach allows you to short Bitcoin (i.e. bet on falling prices) or go "long" (i.e. bet on price appreciation).

---

[46] https://defipulse.com, accessed on 21.09.2021; 15:40.

[47] At the end of September 2020, the market capitalization of all Defis was approximately 19.6 billion USD (defipulse).

[48] https://www.economist.com/briefing/2021/09/18/adventures-in-defi-land, accessed on 21.09.2021; 16:10.

[49] Larcher (2020), p. 98: "An derivative is a financial product that refers to another financial product."

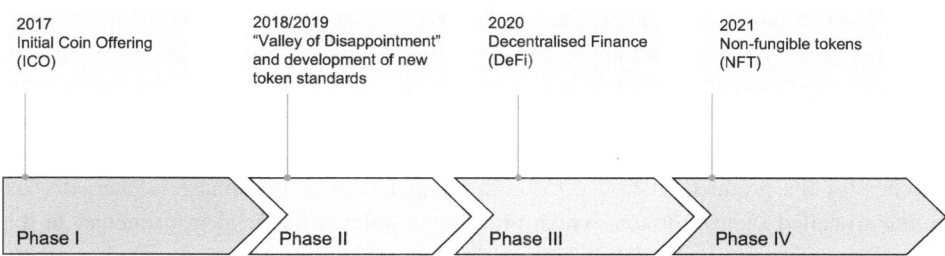

Fig. 6.3 Phases of tokenization

You often hear that Defi will completely revolutionize the financial world. Blockchain technology makes it possible to create these new business models and at the same time make services more efficient, transparent and cost-effective. But a lot of euphoria and visionary business models must not conceal the fact that not all promises can be kept in favor of the users. So in addition to the US regulatory authority SEC, the German authority BaFin is also constantly checking whether the providers are subject to the licensing requirement for banking business. The protection of investors is to be guaranteed as well as the stability and integrity of the financial market.[50] How the regulatory authorities will "intervene" in this market to ensure the aforementioned protective goals is not foreseeable by the end of 2021. The only thing that is certain is that it will be regulated.

Overall, it can be said that slowly but steadily a new economic structure is emerging based on different types of blockchain and corresponding innovative applications. For a long time only ridiculed and only associated with Bitcoin, the realization is now growing in breadth that the blockchain technology creates something meaningful through its community: interoperable, transparent and efficient systems that prevent the concentration of power through the distribution of control over the software. Therefore, it is worth taking a closer look at the technology and the new business models that arise from it.

Figure 6.3 shows the previous development of tokenization.

The first big wave of tokenization starts in 2017, when start-ups use the issue of tokens to avoid the otherwise strictly regulated process of raising capital and, as part of an so-called ICO (Inital Coin Offering), enable investors to participate in growth. For this purpose, tokens of the respective start-up are designed (see ERC20 standard) and paid for with existing crypto-currencies such as Bitcoin. A big mantra of this phase was that now really everyone can get involved early in promising projects—in comparison to the traditional participations, which are reserved for institutional investors in these early phases.[51] However, the vast majority of these projects were not sustainable. This led

---

[50] https://www.fin-law.de/2021/02/08/defi-im-aufwärtstrend-endstation-für-die-finanzmarktregulierung/, accessed on 21.09.2021; 17:30.

[51] Those who can invest in early phases have the chance of big profits if a company is successful. However, the risk of total loss must not be underestimated.

at the latest in the middle of 2018 to the realization among the investors that not every white paper has gone beyond an expression of intent, even if mega-deals were still carried out in the first quarter of 2018.[52]

In the course of ICOs, the focus has been on the tokens and their design. In the first phase, the companies often offered a utility token in their ICOs, which served as a voucher for the product or service to be developed. The second phase is characterized by the so-called security token, which promises regulated financial instruments. In this context, new, innovative models are being devised, which will be summarized under the term "DeFi" from 2020 at the latest. This third phase is not yet complete, but is still developing, also taking into account global regulatory approaches. Since March 2021, even a rather crypto-averse public has become aware that digitally unique works of art can be traded via so-called NFTs. Previously, this was only apparent to the users of CryptoKitties. The next big wave could, inter alia, lie in the combination of blockchain and gaming. In summer 2021, in particular, the company Axie Infinity (AXS) is in the focus of attention. The company enables a combination of blockchain, gaming and NFT and the market data indicate that the "play-to-earn model" used by Axie Infinity is very well received by the users.

6.8   BaFin Regulation[53]

The German Federal Financial Supervisory Authority (BaFin) attributes a high degree of innovation to blockchain technology, which has the potential to influence the financial industry in many ways, e.g. in the area of payments, securities trading, asset management and banking in general.

In this context, BaFin deals with the token economy in order to explain how BaFin as a supervisory authority assesses tokens. The authority emphasizes that it is guided by the principle of technology neutrality in order to uphold the rule-of-law principles of proportionality and equal treatment.

Capital raising through ICOs has disruptive character!

According to BaFin, the classification of a token as a security or financial instrument must be made in accordance with the relevant laws and regulations. European law is to be taken into account. In particular, when designing tokens and their rights, the applicable supervisory law pursuant to the

- Securities Trading Act (WpHG)
- Securities Prospectus Act (WpPG)

---

[52] https://research.tokendata.io/2018/11/07/july-had-the-lowest-ico-activity-in-more-than-a-year/, accessed on 21.09.2021; 21:50.

[53] BaFin, 2018, BaFin Perspectives.

- Investment Act (VermAnlG)
- Capital Investment Act (KAGB)
- Banking Act (KWG)
- Insurance Supervision Act (VAG)
- Payment Services Supervision Act (ZAG)
- Market Abuse Regulation (MAR)
- Directive on Markets in Financial Instruments II (MiFID II)

is to be checked.

MIFID II restricts so-called dark pools. These are private markets on which investors can buy or sell shares and investments. On regulated markets such as the Deutsche Börse, the purchase and sale is carried out publicly and transparently by naming the share package and the purchase price. This transparency is lacking on private markets and thus leads to undesirable asymmetries.

Token issuers must make sure to follow all guidelines and laws. The examination is carried out on a case-by-case basis. However, many tokens are only traded on cryptocurrency trading platforms outside of Europe. If a token is to be classified as an investment fund, the KAGB applies in Germany. If a token is not classified as a security or investment fund, but as an investment within the meaning of the VermAnlG, the provisions of the VermAnlG apply to it.

Therefore, it is necessary to take care of the legal classification of the token. A token can be considered a unit of account. Then MIFID II, WpHG and KWG apply. The supervisory classification of a token as a security does not require physical securitization in a certificate or global certificate. Rather, it is sufficient that the owner of the token can be documented in any case, e.g. by means of blockchain technology.

The guidance letters issued by BaFin are addressed to all market participants who operate in Germany

- Banking business,
- Provide financial services or other services subject to authorization,
- Offer securities or other assets to the public for purchase

without regard to which country the market participant has its registered office. If a market participant addresses its offer to the German market, it is subject to German regulation and the laws and regulations.

Overall, every (potential) market participant is well advised to inform themselves early on in the event of a token offering, which of the laws and regulations are to be observed.

6.9   Conclusion

The dynamics and speed of blockchain development is particularly well illustrated in the field of token economics. With the ability to create decentralized apps on the Ethereum blockchain, new business models have come onto the market almost explosively. This is especially true for startups that want to consistently digitize traditional business processes. For this purpose, startups need appropriate capital. This capital is now no longer made available by the venture capitalists who have been so common for so long, but can now take place in small, but many tranches and worldwide by crowdfunding. This also allows private investors to access highly innovative companies that may have a very high market capitalization in the future.

Both the ICO and any other form of token-generating event take place before a move to the traditional stock exchanges. And usually private investors can only invest in interesting companies after the IPO. Then moderate returns are often to be expected.

Blockchain technology makes more possible. Previous participation and the implementation of new business ideas that might otherwise have been omitted for lack of financing.

Which token model is beneficial for which business idea has to be checked in each individual case. A security token has a great similarity to a security while the utility token is bound to the network of the issuer. In particular, the obligations arising from the issuance of tokens by the issuer must not be underestimated. A request to BaFin is recommended if you plan to issue a token.

References

Altmann J (2003) Volkswirtschaftslehre, Einführende Theorie mit praktischen Bezügen, 6., neu bearb. Aufl. Lucius & Lucius, Stuttgart

Becker HP (2012) Investition und Finanzierung, Grundlagen der betrieblichen Finanzwirtschaft, 5. Aufl. Gabler, Wiesbaden

Berk J, DeMarzo P (2019) Grundlagen der Finanzwirtschaft, Analyse, Entscheidung und Umsetzung, 4., akt. Aufl. Person, Hallbergmoos

Bogensperger A, Zeiselmair A, Hinterstocker M (2018) Die Blockchain Technologie, Chance zur Transformation der Energieversorgung. Forschungsstelle für Energiewirtschaft, München

Bundesanstalt für Finanzdienstleistungen (2018) BaFin Perspektiven. https://www.bafin.de/SharedDocs/Veroeffentlichungen/DE/BaFinPerspektiven/2018/bp_18-1_Beitrag_Fusswinkel.html. Zugegriffen: 4. Sept. 2019

Casey MJ, Vigna P (2018) The truth machine, the blochain and the future of everything. Harper Collins Publishers, London

Chinadaily (2019). http://www.chinadaily.com.cn/a/201909/04/WS5d6eefa1a310cf3e3556985a.html. Zugegriffen: 11. Sept. 2019

Coin Market Cap (2019). https://coinmarketcap.com. Accessed: 16. Oct. 2019

De Filippi P, Wright A (2018) Blockchain and the law, the rule of code. Harvard University Press, Cambridge

European Central Bank (2020) Report on a digital euro. European Central Bank, Frankfurt a. M.

Günderszene (2019). https://www.gruenderszene.de/business/satoshipay-medien-startup-block-chain. Zugegriffen: 14. Okt. 2019

Hahn C, Wons A (2018) Initial Coin Offering (ICO), Unternehmensfinanzierung auf Basis der Blockchain-Technologie. Reihe essentials. Springer Gabler, Wiesbaden

Hellwig D, Karlic G, Huchzermeier A (2020) Entwickeln Sie Ihre eigene Blockchain, Ein praktischer Leitfaden zur Distributed-Ledger-Technologie. Springer Gabler, Berlin

Sveriges Riksbank (2019). https://www.riksbank.se/en-gb/payments%2D%2Dcash/e-krona/. Accessed: 14. Oct. 2019

The Libra Blockchain (2019) White paper. https://developers.libra.org/docs/assets/papers/the-libra-blockchain.pdf. Accessed: 3. Oct. 2019

Vogelsteller F, Buterin V (2015) EIP-20: Token standard. Ethereum Improvement Proposals, no. 20, November 2015. https://eips.ethereum.org/EIPS/eip-20. Accessed: 4. Sept. 2019

Voshmgir S (2019) What is token economy? O'Reilly, Sebastol

Zheng Z, Dai H-N, Wu J (2021) Blockchain intelligence, methods, applications and challenges. Springer Nature, Singapore

Theses for the Future

Abstract

In 2019, when the following theses were written, blockchain technology is no longer in the foreground, and there are voices that no longer attribute this technology to any greater importance because the actual breakthrough in terms of mass application is still pending. However, the formulated theses should encourage discussion. To write off this technology just because the mass breakthrough has not yet occurred would be too short-sighted. Rather, the multi-layeredness should be set out, which calls for dealing with this technology. It is interesting how the view of this technology has changed in a relatively short time.

Meanwhile, blockchain technology is back in the headlines of the press and its potential is being illuminated. Even in these analyses, the focus is still on payment traffic and the transactions associated with it.[1] The undisputed fact remains that this technology ~~will~~ influences our way of dealing with each other, whether on a societal or business level. The way processes are carried out changes through this technology, which, to take up the definition of the concept from the beginning, also enters the meta-level of our actions.

Digitalization is a unique break with an exponential growth of data and a complexity that exceeds our imagination.

For this reason, I have allowed myself to formulate some future theses. It's almost like looking into a crystal ball, and one or the other thesis may develop differently, but these theses should show you again how important it is to deal with blockchain technology as a whole.

---

[1] For example, The Economist: Down the Rabbit Hole; Issue of September 2021.

© The Author(s), under exclusive license to Springer-Verlag GmbH, DE, part of Springer Nature 2022
K. Adam, *Blockchain Technology for Business Processes*,
https://doi.org/10.1007/978-3-662-65818-5_7

In order to be able to classify the scope of this technology more objectively, I conducted two interviews. My interview partners show the importance and the potential contained in this technology from their point of view:

Mathias Goldmann is VP Finance of the blockchain start-up Constellation, which is working on a blockchain version 3.0 in order to overcome the weaknesses of the currently existing solutions. Mathias lives and works in San Francisco (Sect. 7.5).

Axel von Goldbeck, Partner DWF Berlin,[2] shows how small and medium-sized enterprises can use this technology in the context of corporate financing (Sect. 7.6).

The importance of this technology is evident in the strategy paper published by the German government in September 2019. This strategy paper is also referred to.

7.1   Thesis 1: Here to Stay

The trigger for this technical movement was the white paper by Satoshi Nakamoto, who can be described as a highly gifted compiler. Nakamoto takes existing components from cryptography and software programming and brings them together in a way that has not been seen before. This brilliant performance ensures that visions of reality approach. All this happens under great dynamics and yet not in a linear way. This blockchain technology turns our understanding of processes, organizations and patterns upside down. Known roles are discarded, new patterns (e.g. collaboration) arise, and everything is in motion. These upheavals shake our society—and thus trigger emotions.

We live in a complex world that does not always reveal its connections. The political world stage shows us every day how unpredictable our world is. Even if the salvation is not to be sought exclusively in technology or technology, new technologies can give us a promise of order in the existing chaos to a certain extent.

Blockchain technology, referred to by the Economist in 2015 as the "Trust Machine", can provide transparency and reliability. This can restore lost confidence in the economy. Contrary to the impression that the Economist has created, it does not replace human trust with technology. By creating automated processes that are transparent and traceable, the individual does not need to trust that everything will be "good". It is traceable. Therefore, business relationships, as we know them today, are still shaped by human relationships even in the digital age. However, participants in mass business must believe (in the sense of trust) that what is declared is true. In terms of liability, for example, this results in completely new realities for customers, consumers, producers and suppliers. Every food or other consumer scandal can be processed faster and the question of responsibility can be clarified more quickly. The mere prospect of this will lead to more discipline and honesty in everyday economic life and will not require trust in the traditional sense.

---

[2] DWF is an internationally active law firm.

This restructuring will take place in the background with blockchain technology. The individual does not need to be able to program in order to benefit from the advantages.

Currently, the technically oriented discussion dominates, linked to the fear mentioned above, of not being able to understand or use this progress. If you ask yourself why this fear could develop, it also becomes clear that the promise of the Internet, which was also a decentralized network in its origin, was not kept. Even if access to the Internet is possible almost everywhere on this planet, this access does not mean participation in power and decision-making. The fear of being "left behind" is in the room. This results in a distortion of perception and assessment. Some see these and other technologies as "enablers" of completely new approaches, both economically and socially. Others, on the other hand, do not want to be understood as a predictable (human) algorithm and thus deprived of their own individuality. They see themselves and their (working) world threatened by the use of machines and computers.

Regardless of whether the discussion about blockchain technology is emotional or objective, this technology will assert and expand its position. The paradigm shift has begun, even if it is not immediately apparent.

Traditional data-centered business models often depend on a central entity that is equipped with decision-making power and control over all data stored in a particular database. As a result, other parties must simply accept, without specific evidence, that the information transmitted is complete, credible, and accurate, and that the central entity does not use its data for its own benefit.

There are variations; most blockchain solutions allow for the execution of transactions and the shared use of property in a peer-to-peer relationship, with multiple identical copies of the data stored in separate nodes of the network. The blockchain strictly prescribes to data and digital asset owners who and how to access what. The technology's consensus mechanism ensures that these copies cannot be retroactively altered and authenticates the digital assets underlying each transaction. In this way, the blockchain eliminates central entities and serves as a so-called "single source of truth."

Our analog/partially digital world has always consisted of networks. In the Middle Ages, Jacob Fugger was able to build networks: European branches, own courier service, and advantageous contacts with power. To gain allies, make promises, and at least partially keep them, to grant and gain trust, all of these were and are the ingredients for gaining decision-making power in centrally orchestrated societies. Blockchain technology is to be understood in this context as a recipe: The ingredients are still the same—only the environment in which we mix these ingredients anew has changed.

It is not only about an improved, because digitalized network structure with blockchain technology. It is about participation and about the reduction of asymmetries. This reduction leads consistently thought to the (maybe) once in a lifetime opportunity of redistribution. Centralized networks, which accumulate great (data) power and financial resources in order to impose their understanding of economy and order on society, contribute to the division of society into rich and poor. It is not the community that counts, but the centralized holder of power.

It can be rightly objected that also blockchain companies follow this "the-winner-takes-it-all mentality". But if we look at Satoshi Nakamoto's white paper, we overcome this centralist approach for the benefit of the peer-to-peer community. In times of disruptive change, this technology welds people together again to real networks and communities, also and especially to allow and promote market-oriented business models.

Blockchain technology makes it possible to create self-determined and enlightened societies. Through this technique, the society is supported in such a way that transparent and resilient applications arise.

This technology, working in the background, has come to stay!

7.2   These 2: Liberalization of the Internet

As part of the European legislative process, the liberalization of the telecommunications sector began in the mid-1980s. As a result, the German postal monopoly fell, and new providers of telecommunications services, both in terms of equipment and services, entered the market.

In 1989, the Federal Republic of Germany implemented the separation of sovereign and entrepreneurial functions in postal services, resulting in the three companies that still exist today: Telekom, Postbank, and Deutsche Post. However, the separation between "broadcasting and telephone" remains. The European legal framework for electronic communication includes the infrastructure of electronic communication networks—audiovisual media services receive their own directive.

The Internet, as we know it today, started in the 1950s during the Cold War. The American agency Advanced Research Projects Agency (ARPANet for short) was founded to win the race for technical supremacy against the Soviet Union. The ARPANet implemented for this purpose was to be a completely new network that is controlled not centrally, but decentralized, in contrast to the previously usual networks. The network should also continue to function if some locations of the network fail. Four universities that also researched for the project formed the initial ARPANet. The ARPANet made it possible to communicate uniformly over long distances, which was realized by data packets sent over telephone lines. And today our Internet works in principle the same way, although the data transfer speed is many times higher than at the beginning of the ARPANet.

With the beginning of the 21st century, the prerequisite for today's disruptive digital economy has been created, in which the production and distribution of digital content has become comfortable and easy. Digital content in the form of images, videos or blogs is created and published on a large scale today.

The Internet we are familiar with initially gives each individual many possibilities to design their own information supply according to their individual needs. This is "felt" as a direct P2P process and thus understood without the involvement of an intermediary. Everything seems possible on the Internet and the freedom seems unlimited. The dark side of this apparent freedom is known to us as the source of, for example, fake news.

The—at least from the perspective of blockchain technology—legitimate question is whether we do not abuse this freedom to some extent because there are no or hardly any consequences of our actions. We do not consider the price that is paid for this apparent freedom: the surrender of our own data. The big Internet companies build their power on our data. Precht points out in this context that we […] "exchanged autonomy for convenience, freedom for comfort and consideration for happiness."[3] Or, in other words, bread and circuses for the people.

Since no one today has the time to check the authenticity of news or videos shared on social media platforms, for example, it is becoming increasingly important to check the authenticity of the information, i.e. where it comes from and who created it.

With the comprehensible and transparent character of the blockchain, it is possible to verify the authenticity of the information or its sources and to build trust in the news displayed on the Internet. The blockchain in, for example, the news industry makes it possible to produce and distribute content immutably and securely over the Internet.

Blockchain technology makes it possible to strengthen self-responsibility and secure one's own data on the Internet 3.0 and thus also liberalize the economy!

7.3   These 3: Expansion of "Surveillance Capitalism"

This term, coined by former Harvard professor Shoshana Zuboff,[4] describes that the most valuable resource in the age we are moving towards is the prediction of human behavior.

The tech giants like Google/Alphabet, Facebook, Amazon, but also Alibaba, Baidu and Tencent have created quasi-monopolies by repeatedly surprising society and creating facts. The internet companies have managed to convince us that their practices are the inevitable result of the growth of digital technologies.

These companies unilaterally claim our private human experience as their free raw material source. Artificial intelligence is used to predict what the customer wants. We have all learned, starting with Porter's value chain model from 1985, that to lead a successful company, customer needs must be the focus.

We are now on the way to a point where the pendulum swings against us as consumers and customers in favor of this service-oriented thought. It is oriented towards a monopoly position of a few companies. These see in man an algorithm-driven being that is predictable and therefore manipulable in its behavior. The next level in this "prediction game" will no longer be the prediction and preference for certain products, but the recognition of our emotions. Which signs can be recognized and thus exploited before the actual feeling? Which of the approximately 90 head/facial muscles react before the

---

[3] Precht (2018), p. 69.
[4] Zuboff (2019).

feeling enters consciousness? Knowledge of this kind makes a whole different level of predictions possible—from a monopolistic company perspective, this is simply great, because products can be created over the customer—not for the customer.

This form of market economy shows anti-democratic and elitist features that evoke memories of the feudal system of the Middle Ages. In the Middle Ages, the monarch and the nobility formed the authority that one followed, today one follows Google, Amazon & Co. There is a justified fear of being "lulled to sleep". Amazon's prediction algorithm makes such beautiful suggestions about the next book to read, Spotify suggests music titles that match my taste in music and my mood. How convenient if an algorithm tells me what I want next. But with this model we move into minority.

It is up to us to get the pendulum in the center to stop. None of us wants to be so predictable. Autocratic economic systems prevent participation and are striving to dominate a network.

This can be countered with decentralized, distributed networks. Blockchain technology enables a new form of distribution. With blockchain technology, for example, the sovereignty over one's own data can be regained. This would break the business model of the current Internet giants. This resulting vacuum can be divided into small pieces and the dominance in the network can be given back to the actual "raw material suppliers", namely the users.

Perhaps at present not so much points to a system change in favor of a blockchain solution, but revolution and disruption always have something unexpected and usually start quite small in a niche.

7.4   These 4: Protection Against Hacker Attacks

In the age of digitalization, we are connecting more and more technical devices with each other and allowing them to communicate with each other. In 2017, the research company Gartner spoke of the fact that in 2020 about 20 billion devices can interact with each other. However, the computer company Cisco already believed in 2014 that by 2020 about 50 billion devices can be interconnected and thus also communicate with each other.

If our lives take place more and more on our devices, if they perceive and record all the activities we generate, the question arises as to how to protect one's own data.

2018 is a year in which the vulnerability of the general public becomes aware for the first time. Governments, universities, energy suppliers and a variety of companies have become victims of sophisticated hacks. Lack of cyber security threatens personal and corporate data.

Data is a valuable raw material—for a company as well as for hackers. As long as companies use centralized data storage to store their own data, they are highly attractive to attackers, because with a successful attack they can copy or steal all data at once.

Therefore, blockchain-based storage solutions are increasingly being considered as options. Thanks to the decentralized network of the blockchain technology, hackers no longer have a single entry point. Attacking a node within a blockchain network does not allow access to entire data sets.

Access to traditionally secured data sets and database systems often takes place at the devices of the periphery. Routers and switches are examples of this. But other devices such as smart thermostats, doorbells and even security cameras are vulnerable to network attacks. Blockchain technology can be used to protect systems and devices from attacks. The blockchain technology can give these Internet-of-Things (IoT) devices enough "intelligence" to make security decisions without relying on a central authority. For example, devices can form a group consensus about the normal events in a certain network and block all nodes that behave suspiciously.

The blockchain technology can also protect the entire data exchange between IoT devices. It can be used to enable secure data transfers in real time and to ensure timely communication between devices thousands of kilometers apart. In addition, blockchain security means that there is no longer a central entity that controls the network and checks the data passing through the network. Starting an attack is infinitely much more difficult.

The Domain Name System (DNS), whose task is to answer queries for resolving name queries and one of the most important services in IP-based networks[5] represents, is largely centralized. This makes it easier for hackers to break into the connection between website name and IP address and cause chaos. They can seize websites, lead people to fraud sites, or simply make a website unavailable. They can also DNS attacks with DDoS attacks[6] combine to make websites completely unusable for a longer period of time. The currently most effective solution to such problems is to track log files and enable real-time alerts for suspicious activity. A blockchain-based system as a decentralized network makes it much more difficult for a hacker to find and exploit individual vulnerabilities. Domain information can be stored immutably in a distributed ledger, and the connection can be made through immutable smart contracts.

Therefore, blockchain is clearly developing into a very practicable technology when it comes to protecting businesses and other network participants from cyber attacks.

---

[5] IP stands for Internet Protocol; it is a widely used network protocol and the basis of the Internet.

[6] DDoS stands for Distributed Denial of Service; in information technology, Denial of Service (DoS) is the unavailability of an Internet service that should be available. This can happen by overloading the data network or by a targeted attack on a server. If these requests, which overload a server, come from a variety of devices, one speaks of a DDoS. Since in a DDoS attack the requests come from a variety of sources, it is not possible to block the attacker without shutting down the entire network.

7.5   These 5: Future on a Higher Logical Abstraction Level (Interview with Mathias Goldmann)

What does Blockchain Technology Mean for you?
Blockchain technology is for me a socio-technical movement that requires a different way of thinking and organizing from us. It requires a paradigm shift and an opening of thinking that will last for many years. We are at the beginning of an extremely interesting development that has great potential and great risks if we fail to actively and value-oriented shape our future.

In What Social Context do you see this Technology?
When Bob Dylan wrote "The Times They are a-Changin" in 1964, a far-reaching geo-political movement was going through the world. Today, these words could not be more current. The times of change are clearly visible everywhere. Every aspect of life is touched by new technologies and the paradigms that come with them. Unlike iterative innovations, truly groundbreaking inventions are characterized by the fact that they permeate our lives at all levels. They touch the social, economic, technological, environmental and psychological fabric of our societies. There is no doubt that the blockchain is such a technology. After the first ten years of its existence, it is time to take a step back and see where the journey is going.

How Would you Describe the Development of this Technology so Far?
The first ten years and three red pills
We have only been ten short and intense years in the development of blockchain technology. At the same time, the technology has undergone three iterations in its short existence. Each phase brought its own focus, mindset, application areas and business applications with it. Each development phase corresponds to a "red pill moment" as it is known from the film "The Matrix". Neo played by Keanu Reeves opens up a completely new reality by swallowing the red pill offered by Morpheus (Lawrence Fishburne). From this moment on, there is no turning back. Let's follow in the footsteps of this evolution to describe the future.
The first red pill: Bitcoin & the global API for value
Satoshi's whitepaper was published in 2008. It was the beginning of a movement and a completely new way of thinking. It was so new that many people are still trying to understand this paradigm in its entirety even after 10 years. Born in the financial crisis of 2008/2009, Bitcoin is widely seen as an ideological-technical response to the failure of the debt-based, compound interest and fractional reserve banking and money system.

While money can be seen as an API for value (a universal interface that allows the exchange of various things with each other), Bitcoin can be understood as a global API to transfer value safely and autonomously.

The second red pill: Ethereum & smart contracts

While the monetary implications of Bitcoin received global attention, the technology on which Bitcoin worked remained largely unnoticed.

In the second "red pill moment" in the history of blockchain, the Ethereum network went live in 2013. Based on the concept of "smart contracts", it not only allowed the transfer of value, but also of information. Smart contracts are self-executing programmable logics in which the conditions for a value or information exchange are inevitably set.

Ethereum brought blockchain technology into the spotlight of attention. At the same time, it launched the ICO wave of 2017 with the ERC20 token standard.

A completely new ecosystem with a current market capitalization of 50 billion USD resulted. While in 2017 everything seemed ripe for tokenization, the bubble collapsed mainly due to speculation due to a combination of technical, regulatory and timing aspects. While it has become much quieter in the blockchain industry, we are moving towards the next "Red Pill Moment".

The third red pill: DAG & Data Economies

From a technical point of view, it has become clear over the years that classical blockchains such as Ethereum have a fundamental conceptual bottleneck: applications running on such a network share the complete data throughput of the entire network with each other. This means that an application with high data throughput requirements blocks the network, and transaction costs rise very high. This and other reasons such as governance are the reason why the Ethereum network never really came into question as a global infrastructure for companies.[7]

The solution lies in the third generation of blockchain solutions and thus describes the third "Red Pill Moment": Blockchains of the latest generation are horizontally scalable and based on a network structure that is called DAG (directed acyclical graph). Current and serious examples are the companies COTI, Constellation, Scroll Network and a handful of other companies. A DAG graph-based network allows the asynchronous processing of transactions in the network. At the same time, the scalability ensures that the data throughput rate increases with the number of network participants! Due to the technical innovations, the focus for blockchain applications shifts from smart contract logics to high data throughput rates and thus "enterprise readiness". This opens up completely new application areas and future scenarios that are data-based.

What Potential do you see in the Future Application of this Technology?

The infrastructure of the third generation makes Big Data and data stream processing possible on a blockchain. A high data throughput rate enables the notarization of data-rich videos, sensors or machine learning pipelines. Big Data plus Blockchain thus becomes reality.

---

[7] State-channel solutions such as Plasma and Raiden are not (yet?) functional for record-keeping.

This alone brings dozens, if not hundreds, of industrial applications to mind. Applications in which a blockchain brings advantages through interoperability, notarization, audit trails and encryption.

Data sovereignty

The problem with data today is that those who generate the data have no control over the data. Third parties have built a billion-dollar industry on information they have collected from others. Many people see this as a form of theft and violation of their privacy. They feel that they should have full control over their own data. The third generation of blockchain technology allows the data monopoly to be returned to the "producer". More than that, data sovereignty allows the producer to monetize his/her data himself/herself. The principle is equally applicable on an individual and on a corporate level.

Data sets can be made available on the network without violating regulations. This is possible by comparing and evaluating hashed metadata. Furthermore, validated data packages can be tokenized in the network. This allows individual data packages to be assigned a value. This makes the network infrastructure into an API that can connect various data sets while at the same time creating a data marketplace in the network itself.

In such a marketplace, a data economy arises in which those who generate the data have full control and direct monetization opportunities without intermediaries.

Data Economies & data-based entrepreneurship

When sovereign actors develop data economies around data, this creates a new business activity: data-based entrepreneurship.

These entrepreneurs or companies sell validated datasets for money or share and exchange them with other datasets that, for example, need to be available cross-border for the operation of new technologies (for example, autonomous vehicles). The entrepreneur can tap into entire data silos that are unused and available in his/her company, an industry, or even a country and monetize them. This will trigger a second data-based gold rush, in which blockchain technology provides the infrastructure to validate, share, and sell data securely. All of this happens while respecting privacy and regulations.

In this future scenario, the understanding of business models will increasingly shift from the physical to the digital. The physical product will become less important, while the digital metadata and properties around the product will become increasingly important. Automobile companies understand this change. They have recognized that the image of their brand will depend less on the physical product than on the services around the mobility service. Data products and their monetization will therefore play a major role in a future in which the physical product will become almost unimportant.

Digital twins & the solution to the world's most pressing problems

Resource depletion, waste, environmental pollution, habitat destruction are directly linked to too much economic activity and consumption. Matterum is working on a groundbreaking solution to address the systemic problems associated with our economic production system. By combining digital twins with blockchain technology, it will be possible to usher in the next industrial revolution and find new ways out of the global systemic impasse.

A digital twin is a data set that contains information about the properties and life cycle of a physical object. Digital twins enable a completely new way to produce, trade and own physical objects. Why? A digital twin is an identity card that accompanies the entire life cycle of an object. This means that the state, resources and status of the object are available as information. This allows for more accurate and intelligent planning of resources and production. It opens up secondary markets for used and older objects that would otherwise have no liquidity. A universal naming system for physical objects also allows for the efficient exchange (keyword: data economy) of object-related meta-information (quality, condition, location, age, etc.).

In this sense, the entire life cycle of an object is planned. Higher quality and lifespan are incentivized, and inefficient cheap production for the landfill is reduced.

Global sovereignty, mindset & governance

As Albert Einstein already said: "Problems can never be solved with the same mindset that created them."

Today it is more than ever evident that humanity needs a structural change in its mindset and social organization to meet the immense problems of global proportions. In this sense, selfish and isolationist thinking and behavior as well as classical control and organization structures have served their purpose.[8] Fortunately, blockchain technology as a fast, global and systemic technology promotes the mindset, logic and organizational structure to address these challenges.

One of the unique phenomena is that technology has resulted in the "birth" of global communities. The support of blockchain companies by these communities is essential for success. The networking, speed, ingenuity and financial clout are present to a degree never seen before.[9]

In the future, the global networking of communities with a common goal will increase. The baby boomer generation will have stepped down from the levers of power in the next 10 years. This will make room for systemic thinking, which, supported by blockchain technology (among others), penetrates all areas of society.

---

[8] Frederic Laloux's book "Reinventing Organizations" lies at the intersection of decentralized networks, economy and social organization. Laloux is a Canadian researcher who has studied companies around the world that functioned not through classical top-down hierarchies, but through self-organization. Against this background, decentralized networks fit very well into the paradigm of new social self-organization. Interestingly, the book was published shortly after Ethereum went live.

[9] From a historical perspective, most social movements have always struggled for financing. Global blockchain communities differ significantly from this. In addition to human and brain capital, they also have excellent financial resources to support ideas. While early blockchain visionaries are highly idealistic and futuristic, at the same time, much potential was lost in the speculation bubble of 2017. Nevertheless, the global community aspect still exists and will be back in the spotlight with third generation blockchains.

Systemic thinking will deeply penetrate everyday life and human interaction. It will force people to understand themselves as a node in the entire social fabric (= network). Nodes that act against the common good of the whole will be excluded from the network and its resources, similar to a blockchain network. A positive global future will simply not be able to afford the old patterns of behavior and ideas.[10]

This means we have an answer to Carl Sagan's question: "Who speaks for the earth?"

We answer: Systemic thinking that sees the earth as a sovereign entity that represents the highest level of global governance.

This means from a governance perspective that decentralized networks form a new global entity that goes hand in hand with our thinking models.

In the future, these networks will have a new legal status. Through new governance mechanisms, they will enable participants and stakeholders to be better represented than today's systems.

In this sense, decentralized networks can be most compared to organizations like the United Nations. They force us to think globally and to abandon the silos of nation-state, territory, and citizenship in favor of global sovereignty and governance.

What is your Recommendation in Terms of Blockchain Technology?

Anyone who deals with blockchain technology must take into account the larger social context. Technology does not arise in a vacuum, and at the same time technological innovation drives social change. This is unavoidable. As individuals and society, we need to zoom out and locate social change in its historical context in order to develop a positive and realistic future. Roughly sketched, Internet technologies have resulted in a worldwide networking and greater exchange of information and more opportunities. The trend is from individualism and isolated views to a globally connected systemic approach in which we are all part of the great system of Earth.

Blockchain is no exception as an extension of network-based technology. With this global networking comes a structural change from top-down hierarchies to self-organized clusters and networks. I recommend that everyone interested read the book by Frederic Laloux "Reinventing Organizations". It describes how economic success is achieved under a completely different paradigm of self-organization and what challenges this poses for individuals and companies.

Well, no change comes without risks and without price.

The paradigm shift around blockchain technologies requires us to be more awake and louder in demanding our values.

It requires us to actively shape the future, rather than wait for regulators and legislators. It requires an active understanding of democracy within a global society.

It requires active participation in shaping our future as we see fit.

---

[10] Power, religion, race, environment, social organization, etc.

This means that many of the local ideas about identity, nation-state, and economic best practices must be thoroughly questioned and emotionally experienced in order to lead to a new understanding of being.

If we fail to do this, the otherwise neutral technology can become a "dictator's wet dream."

Therefore, my recommendation is a deep engagement with the topic in relation to the biggest questions and challenges of our time.

7.6   These 6: Blockchain and Mittelstandsfinanzierung (Interiew with Axel von Goldbeck)

What does Blockchain Technology Mean to you?
So far, blockchain technology has yet to fulfill the great promises that are associated with it. From my point of view, however, it is already foreseeable that it has considerable potential for value creation. Otherwise, the considerable investments made by companies of the "old economy" cannot be explained. On the other hand, the business is partly driven by ideology. Blockchain technology is clever, but not "better" or "worse" than other technologies. I consider it naive to expect a new, more democratic world from it. Currently, it is all about passing the practical test. And that is being undertaken by the conventional players according to classical cost-benefit considerations. I secretly hope that one or the other malpractice in our economic system can be rectified with the blockchain.

In What Social Context do you see this Technology?
In small and medium-sized enterprise financing, it has been a financial policy issue for years: in many countries, not just in Germany, small and medium-sized enterprises contribute significantly to economic value creation. And creates more jobs than industry. And yet, studies are increasing that the financing situation of small and medium-sized enterprises is becoming increasingly severe. "SMEs currently make little or no use of alternative financing to traditional bank loans—the house bank remains the most important financial partner."[11] At the same time, "the constantly tightening equity requirements for banks and the disproportionate lending from the past lead to a restricted or delayed lending readiness for SMEs."

While large companies regularly and increasingly finance themselves through the capital market, small and medium-sized enterprises rarely resort to capital market financing. The reasons for this are manifold. The main ones include the costs of a capital market financing, which usually eat up around 4% of the emission volume. Other reasons also

---

[11] Deloitte, Study Financing in the Mittelstand—Optimization Potential in Mittelstand Financial Management, 2019.

play a role: SMEs often shy away from the effort involved in informing investors or the publicity obligations associated with capital market activities.

The increasing digitalization contributes to reducing the barriers to capital market financing. Numbers are held and evaluated digitally. Professional ERP systems make it easier to control where they are used.

The blockchain technology (blockchain and distributed ledger technology are used synonymously here) has the potential to further simplify capital market financing. The tokenization of bonds will soon take on industrial forms. More and more projects are being carried out by institutional issuers and companies. Examples include Daimler's and LBBW's large-scale issuance of bonds based on blockchain technology. So-called security token offerings are spreading to more and more areas: from venture capital financing to financing of real estate and other tangible assets, so-called asset token emissions and fund financing.

To avoid misunderstandings, blockchain-based financing is not a different type of financing. There is indeed a blockchain-based financing form next to the classical financial instruments in the form of so-called utility tokens, which represent a kind of digital voucher. But here too, only existing legal instruments (vouchers) are "tokenized", i.e. converted into digital representatives. The same applies in the classical corporate financing area. Here too, only digital representatives of conventional financial instruments are created, which can be registered and traded in a distributed register. This focus on the "digital ledger" in connection with smart contracts simplifies the emission and trading process considerably—not least because the emission and trading process previously split into numerous services is merged into one register. This is where the real efficiency gains lie.

How Would you Describe the Development of this Technology so Far?

The technology is going through the classic boom-and-burst development, from a payment application to the ICO hype to classical financing and many other, more useful applications. From my point of view, the development is something logical. I'm glad that the post-ICO crisis didn't last that long, but blockchain applications are increasingly being adapted and found useful by rational players.

What Potential do you see in Future Applications of this Technology?

The exciting thing about the blockchain is that it is not yet clear in which areas it will develop. This will still be an adventure, with all the chances and risks involved.

One of the not-to-be-underestimated advantages in my area of financing lies in the fact that "tokenization" makes previously illiquid values liquid. This can best be illustrated by the well-known security token emissions in Germany: Although often referred to as a bond or—equivalently—a loan, it is actually a contractual agreement. This can be easily explained by the fact that bonds in Germany have so far been subject to the formal requirements of a written contract. A bond without paper is therefore not currently (November 2019) possible in Germany. If you look at the terms and conditions of these instruments, you will find that the transfer is "contractually" regulated, i.e. by assignment or contract takeover. The paper, if any, follows the law. A bond, on the other hand, is transferred according to property law principles: The law follows the paper. This greatly simplified the transfer in the paper age. For bonds, the Federal Government plans to abolish the requirement for a written contract. This was supposed to be a draft bill in 2019. Now it is expected that this draft will be completed in the 2nd quarter of 2020.

The importance of these questions, however technical or legal they may sound, is often underestimated. Drastically reduced emission costs make it easier for medium-sized companies to access the capital market. On the other hand, the tradeability of previously illiquid instruments makes it easier for investors, small and large, to take on the risk of financing small and medium-sized enterprises.

In addition, in view of the European-wide financing problems, the European legislator has taken some steps to limit the information obligations of securities issuers to an appropriate level. The fulfilment of information obligations is one of the largest cost blocks. In view of five- or often six-figure amounts incurred in this context in terms of consultant costs, smaller companies with low emission volumes shy away from capital market emissions.

The EU has now taken this into account with a revision of the Prospectus Directive in 2017, the final stage of which came into force in July 2019. Public offers with a total volume of up to 8 million euros p.a. are no longer subject to the prospectus requirement. Member States may guarantee investor protection in other ways for offers from 100,000 euros. The Federal Republic of Germany has availed itself of this by introducing a securities information sheet (WIB) for public offers of securities, which is to be approved by BaFin. The maximum scope of the WIB is limited to three pages. To compensate for the allegedly lower investor protection of the WIB, securities with a total volume of 1 million euros to 8 million euros p.a. may only be intermediated by investment advisers, investment intermediaries or a securities service provider that is legally obliged to check whether the total amount of securities that can be acquired by a non-qualified investor does not exceed certain maximum investment amounts for this investor group. In addition, there are further simplifications for so-called growth companies through the revised EU Prospectus Regulation.

This may all sound quite complicated. However, in sum, public offers on the blockchain represent a significant simplification for capital market financing of medium-sized companies. The amounts that can be collected annually without a prospectus represent decent amounts for many companies. Securities tokens are not yet a common financial instrument. Classical investment advisers, investment intermediaries or securities trading companies are only occasionally found in this market. But if more investors learn to appreciate the liquidity of digital securities tokens, these professions will not close themselves off to new technology. First emission platforms that want to take on the role of banks for these emissions are applying for investment intermediation licences.

In addition to conventional financial instruments, tokens offer further opportunities to finance illiquid values quickly. The potential of utility tokens has not yet been exhausted.

The great flexibility in the design of tokens creates further innovative financing methods. This makes financing more diverse and probably also more opaque. But those who learn to "play" the capital market and investor needs will no longer be hindered by high costs and complex procedures.

7.7   These 7: "Regulatory is in" or Blockchain Strategy of the Federal Government

On September 18, 2019, the Federal Government presented and adopted its strategy on blockchain technology in the Federal Cabinet.[12] The Federal Government also sees in this technology "the building block for the future of the Internet". 44 measures are proposed to better assess the technology with its opportunities but also risks. It is explicitly stated that the Federal Government is challenged to contribute "to the clarification and exploitation of the potential of blockchain technology as well as to the prevention of abuse possibilities".

The vision of the Federal Government is to not only maintain but also expand Germany's leading position in this technology with its many facets of application. Germany also has a good reputation internationally with regard to blockchain competence and community. So many developers in Germany are working on further developing the infrastructure and implementing useful applications. This can succeed if the necessary space for experimentation is granted.

Therefore, the federal government has addressed five areas of action:

- Secure stability and stimulate innovation: blockchain in the financial sector.
- Maturing innovations: promotion of projects and real laboratories.
- Enable investments: clear, reliable framework conditions.
- Apply technology: digital administrative services.
- Spread information: knowledge, networking and cooperation.

The focus is initially on the financial sector, as the first application case in the form of the digital currency Bitcoin has from the outset inspired the imagination of developers and blockchain enthusiasts. In 2019, the federal government began work on a draft law that is to allow bonds to be represented by a blockchain solution. To date, late spring 2020, this draft has not yet been published or submitted to the Bundesrat for comment. The next step is to examine how investment fund shares and digital shares can be mapped using blockchain solutions. ICO, token trading and electronic securities are thus

---

[12] Federal Government's Blockchain Strategy (2019), https://www.bmwi.de/Redaktion/DE/Publikationen/Digitale-Welt/blockchain-strategie.html; accessed: September 18, 2019.

regulated by law. New legal regulations develop standards that enable previously necessary physical templates in paper form to be replaced by digital files.

Further exploration of value-added application possibilities is to take place, inter alia, through so-called real laboratories. The findings from these laboratories are to serve as preparation for mass-compatible applications.

It is interesting that the federal government takes a position on this technology. But: Germany as a business location is also under pressure. As already reported in sect. 6.2, the Principality of Liechtenstein has passed a blockchain law. Germany is interested.

As much as the activities to make Germany one of the leading blockchain nations are to be supported, so much attention must be paid to the fact that not too much regulation takes place. Although it is criticized—also on the part of the blockchain community—that a lack of legal framework drives innovation abroad, Germany has had very good laws for a long time, which were a model in other countries time and again. The existing legislation must be transferred to the digital age and overall lead to a reduction in bureaucracy!

References

Bundesregierung (2019) Blockchainstrategie der Bundesregierung. https://www.bmwi.de/Redaktion/DE/Publikationen/Digitale-Welt/blockchain-strategie.html. Zugegriffen: 18. Sept. 2019
Precht RD (2018) Jäger, Hirten, Kritiker. Eine Utopie für die digitale Gesellschaft. Goldmann, München
The Economist (2021) Down the rabbit hole. Ausgabe 18th September. The Economist
Zuboff S (2019) Das Zeitalter des Überwachungskapitalismus. Campus, Frankfurt a. M.

Outlook

8

Abstract

This book discussed the different areas of use for this technology. This chapter summarizes the findings and explains the advantages and disadvantages of this technology, and invites participation.

The blockchain technology, as well as the technique, fascinate. The possibilities seem endless. But at the latest when you think about using this technology in your organization (whether it is a company or an authority), you notice the complexity of this technology, based on the diverse possibilities.

In order to grasp the potential, you need more knowledge than just the knowledge that Bitcoin is a digital currency to which a rather dubious reputation of speculation is often attached.

Therefore, it is important to deal with the building blocks of a blockchain and their design options, even if only to reject this approach at the end if it is found that this solution does not provide the desired added value. However, an attitude of rejection due to lack of knowledge is irresponsible. It is important to be able to distinguish the signals from the general noise of information available on this technology. Only in this way can the potential for own solutions be recognized.

Even if it is still a relatively young technology and there are still many questions (asked as well as unasked) in the room, you do not have to reinvent the wheel. There are blockchains! On these blockchains you can test your own ideas. Allow yourself your own "playground" and use the many possibilities. With each run your knowledge and confidence grow, which process you want to optimize and can blockchain-based. For a professional implementation, it may be advisable to involve professionals.

© The Author(s), under exclusive license to Springer-Verlag GmbH, DE, part of Springer Nature 2022
K. Adam, *Blockchain Technology for Business Processes*,
https://doi.org/10.1007/978-3-662-65818-5_8

Let your curiosity be aroused by this technology—and stay curious. That there is more behind this technology than some skeptics want to believe in is at least recognizable by the government's strategy paper. A rather slow and politically democratic body reaffirms the importance of this technology and commits itself to far-reaching promotion. Every company should take this as an invitation to check for itself whether and to what extent it can integrate this technology into its own processes.

For me, dealing with this technology means daily (joyful) learning. The dynamics with which the technology and thus the technology develop is high. The technology requires an interdisciplinary approach. It is not enough to merely check what is technically possible. This technology affects companies as well as society.

From my point of view, it is also important that this technology is one that works in the background. Although a blockchain solution will also have to be connected to a so-called front-end (otherwise how will one access this database), a blockchain unfolds its effect in the background. It protects the data from changes and manipulation attempts. The distributed database structure allows for a very high level of security. All of this will not interest the user of a chic application in the event of doubt. This application can initially offer the user a completely different value, for example, a more comfortable process structure. The optimization of the process structure also depends on the guarantee of data security. Middlemen who have previously guaranteed this by law are eliminated and thus contribute to the efficiency increase within the process. All of this happens in the background. For customers and users, in addition to slim, comfortable and easy processes are important.

In summary, it can be said that blockchain technology (in a broad sense) allows the following:

- It can store data securely and reliably on a blockchain.
- It can execute the most diverse applications on a blockchain.
- The applications you have selected can be designed transparently and logically comprehensible by the blockchain concept.
- By your application you can enable that different parties with different interests can simply, securely and efficiently exchange data and interact with each other very quickly, without network messages being necessary.
- You can build an algorithm into your applications that describes governance rules, e.g. about the procedure for changes in the protocol, in order to be able to react flexibly to future challenges without knowing them in detail today.
- The applications you have developed can still function in the network even if you should lose interest in the application as initiator. This ensures long-term availability that can convince users to trust your application.

In addition to the things that a blockchain can make possible, the question arises as to what challenges lie ahead of this technology:

- The understanding and knowledge of this technology is often highlighted as expert knowledge that is hardly accessible to the general public.
- Blockchain technology is still equated with the cryptocurrency Bitcoin ten years after its introduction. However, this technology can enable much more than just serving as a means of payment.
- In 2019, when this book was first time published, the transactions per second (txs) on a blockchain are not yet high enough to make them a real competitor to traditional databases. Even now, in 2022 this can be improved but it is at the same time knowledgeable that there are providers who can provide up to 100.000 txs. Sounds great and we can see already the improvement, but there is more in!
- The problem of interoperability, i.e. that different types of blockchain can communicate with each other, has not yet been satisfactorily solved.
- Types of blockchain that require authorization are tend to be efficient and lean, but they are centrally structured and organized. This can store data reliably and non-manipulatively, but the original idea of a decentralized and distributed peer-to-peer network is thereby weakened.

  This book shows a small, but very practice-oriented section from the large world of blockchains, which in turn is only one facet of digitalization. However, I consider this facet to be decisive, because although blockchain technology includes a technology that acts in the background, it enables it in the overlapping context with other technologies and technologies to create considerable added value.

To explore this added value for your own company or your own pilot project, the focus of this book is on checking the process and business logic of a potential application case. The workshop steps listed in Chap. 4 have been tested on several student courses. The findings and results that my students have always come to have convinced me to make this approach accessible to other groups of people. It is not just about realigning a process. Rather, the dependencies should be recognized. If they are visible and conscious, then traditional patterns of thought can be broken.

Have fun!

Glossary

51% attack An attack on the blockchain, resulting in a group of miners controlling more than 50% of the network's hashrate; this term is mainly used in reference to Bitcoin.

Arbitrage Is the exploitation of price differences on different stock exchanges and markets at the same time in order to be able to take profits.

Adress In order for participants in a network to be able to carry out transactions of any kind, an alphanumeric address is required. A Bitcoin address, for example, looks like this: 1DSrfJdB2AnWaFNgSbv3MZC2m74996JafV. It consists of a series of letters and numbers starting with a "1" (number one). Just as you ask others to send an email to your email address, you ask others to send you Bitcoin to your Bitcoin address.

Airdrop Giving away tokens to certain blockchain addresses, combined with the intention of increasing interest in a token and a community and thus also increasing the token value.

Altcoin Altcoin is simply any digital currency alternative to Bitcoin. Many altcoins are branches of Bitcoin with minor changes (e.g. Litecoin).

API Stands for *Application Programming Interface*, a software intermediary that helps two separate applications communicate with each other. They define communication methods between different components.

Confirmation As soon as a transaction is added to a block, it receives a confirmation. As soon as another block is mined on the same blockchain, the transaction has two confirmations, and so on. Six or more confirmations are considered sufficient proof that a transaction cannot be reversed.

Bitcoin The name of the currency unit (the coin), the network and the software.

Block A block contains fields that are often self-explanatory: On the one hand there is pure "info data", on the other hand the hashes. Block information includes data such as creation date, size or number of transactions.

Block Explorer This is a website or programme that allows users to search the blocks of a blockchain. It is comparable to the files or folders listed in a computer's explorer.

Block Reward The creation of blocks is rewarded.

© The Editor(s) (if applicable) and The Author(s), under exclusive license to Springer-Verlag GmbH, DE, part of Springer Nature 2022
K. Adam, *Blockchain Technology for Business Processes,*
https://doi.org/10.1007/978-3-662-65818-5

Blockchain A grouping of transactions marked with a timestamp and a "fingerprint" of the previous block. The block header is hashed to create a proof of work to validate the transactions. Valid blocks are added to the main blockchain by network consensus.

Bounty Program Crypto bounty programmes are lists of tasks that generally anyone can participate in and receive tokens from the project. Tasks usually include actions to help grow the community, such as joining the Telegram channel, retweeting content, liking on Facebook and signature campaigns on BitcoinTalk.

Byzantine Generals Problem A reliable computer system must be able to cope with the failure of one or more of its components. A failed component may exhibit a type of behaviour that is often overlooked—namely, sending conflicting information to different parts of the system. The problem of coping with this kind of failure is expressed abstractly as the problem of the Byzantine generals.

Cold Storage Refers to holding a reserve of Bitcoin offline. Cold storage is achieved when private Bitcoin keys are created and stored in a secure offline environment. In contrast, hot storage refers to connecting to the internet. A wallet is hot when it is directly accessible via the internet or resides on a computer with an internet connection.

Colored Coins An open-source Bitcoin 2.0 protocol that allows developers to create digital assets on the Bitcoin blockchain by using its functionalities outside the currency.

DAO Digital *Decentralised Autonomous Organisation (DAO). The DAO served as a form of investor-driven venture capital fund that aimed to provide companies with new decentralised business models. The DAO's code, built on the Ethereum blockchain, was open-source. The organisation set the record for the most crowdfunded project in 2016, but those funds were partially stolen by hackers, prompting Ethereum to create a new blockchain fork called Ethereum Classic, in which all investors were placed as they were before the attack.*

Dapp A decentralised application (Dapp) is an application that is open-source, operates autonomously, stores its data on a blockchain, offers incentives in the form of cryptographic tokens and operates according to a protocol that provides proof of value.

Decentralisation Dispersion of decision-making powers.

Digital signature A digital signature, generated by public key encryption, is a code attached to an electronically transmitted document to verify its content.

Distributed-Ledger-Technologie (DLT) DLT stands as a term for blockchain technology, as it is also a decentralised database.

Dogecoin Dogecoin is a dog-themed cryptocurrency that was launched in 2013 and is an alternative to more well-known currencies such as Bitcoin. Although the value of a single Dogecoin is very small (often part of a cent), the enormous number of Dogecoins in circulation equates to a market capitalisation of over $27 billion.

Double Spending Double spending is the result of successfully spending money more than once. Bitcoin prevents double spending by checking each transaction that is added to the blockchain to ensure that the inputs to the transaction have not been spent before.

ECDSA (Elliptic Curve Digital Signature Algorithm) Elliptic Curve Digital Signature Algorithm is a cryptographic algorithm used by Bitcoin to ensure that funds can only be spent by their rightful owners.

One-way property Two inputs mapped to the same output hash. While hash collisions are possible, it is nearly impossible to provide two meaningful data sets whose hashes collide. Hashes are one-way streets; they can be constructed from data, but data cannot be reconstructed from hashes.

ERC20 Token Standard ERC stands for "Ethereum request for Comment" and the ERC20 standard enables the creation of tokens on the Ethereum blockchain. The smart contract functions used in this standard allow tokens to be issued and their supply, circulation and balance to be monitored. The ERC20 standard has since been extended to meet the requirements of the ongoing tokenisation.

Ethereum Ethereum is an open-source distributed system that offers the creation, management and execution of decentralised programmes or contracts (smart contracts) in its own blockchain.

Ethereum Virtuelle Machine (EVM) A simulated state machine that uses eWASM bytecode to process transactions and perform state transitions for the Ethereum blockchain. Operation is guaranteed, i.e. for each block, the state of the EVM is exactly the same on every node in the network, and it is impossible to create a different state with the same inputs.

EWASM A web assembly (WASM) version implemented by the Ethereum Virtual Machine that provides additional blockchain functionality.

Exchange A service for trading cryptocurrency tokens for other tokens or fiat. Exchanges are one of the few ways to change cryptocurrencies into fiat and transfer that value to a bank account.

Fork A fork, also known as an accidental fork, occurs when two or more miners find blocks at almost the same time. Can also happen as part of an attack.

Fee Sending a transaction often involves a fee to the network for processing the requested transaction. Most transactions require a minimum fee of 0.5 mBTC.

Genesis Block This is the first block of a blockchain. Only when this is present can the blockchain start. Usually, the first block is created manually and added to the software.

GUI *Graphical User Interface* is a way to display information to the user via stylised on-screen elements such as windows and taskbars.

Hash A digital fingerprint of a binary input.

Hash-Function A hash function is an algorithm that uniquely maps a digital input of any length to an output length that is always the same.

Immutability See immutability.

Consensus When multiple nodes, usually most nodes in the network, all have the same blocks in their locally validated blockchain.

Cryptocurrency Digital currencies created on cryptographic tools such as blockchains and digital signature. They are not recognised as currency because they do not fulfil the typical monetary functions.

KYC *Know your Customer* or Know your Customer (KYC) is the process of a company identifying and verifying the identity of its customers. The term is also used for the banking regulation that governs these activities.

Market capitalisation Market capitalisation reflects the current stock market value of a company and is calculated as follows: Shares of the company multiplied by the listed price per share.

Mempool The Bitcoin mempool (memory pool) is a collection of all transaction data in a block that has been verified by Bitcoin nodes but not yet confirmed.

Mining Mining is the performance of mathematical calculation by computer hardware to validate Bitcoin transactions. It is a highly specialised and competitive market where only those who manage to solve the puzzle first reap the rewards.

Mining Pool DThis is a mining approach where multiple generating customers contribute to the generation of a block and then split the block premium according to the computing power contributed.

Network A peer-to-peer (P2P) network that forwards transactions and blocks to every Bitcoin node on the network.

NFT A Non-Fungible Token is a representative of a unique, non-exchangeable or copyable digital asset.

Nonce The "nonce" in a Bitcoin block is a 32-bit (4-bit) field whose value is set so that the hash of the block contains a pass of leading zeros. The rest of the fields must not be changed as they have a defined meaning.

Node A node is an entity in the blockchain network that either proves (public transactions) or validates (hybrid or private transactions) and then adds them to a block with a unique hash. The hash is used as input by the next transaction. Nodes receive rewards when they fulfil their tasks within the network and they thus contribute to blocks with transactions being bound to the predecessor block.

Off-Chain-Transactions An off-chain transaction is the movement of value outside the blockchain. While an on-chain transaction—usually referred to simply as a transaction—modifies and depends on the blockchain to determine its validity, an off-chain transaction relies on other methods to record and validate the transaction.

One-Time-Password A one-time password (OTP) is a type of password that can only be used once. It is a secure way to gain access to an application or perform a transaction only once. The password becomes invalid after it has been used and cannot be used again.

Polkadot Polkadot is a sharded blockchain protocol that allows multiple blockchains to communicate with each other and collaborate efficiently so that they can share heavy workloads and avoid bottlenecks. At its core, the Polkadot protocol is a translation architecture that allows users to combine, decentralise and scale blockchains as needed.

Satoshi The smallest unit of a bitcoin: 0.00000001 BCT.

Satoshi Nakamoto Author of the white paper "Bitcoin: A Peer-to-Peer Electronic Cash System". This author developed Bitcoin and created the original reference implementation, Bitcoin Core. As part of the implementation, he also developed the first blockchain database. In the process, he was the first to solve the problem of double spending for the digital currency. However, the true identity of the author remains unknown.

Level of difficulty The difficulty level is used in the mining of Bitcoins; it is adjusted so that the time for block detection can be kept constant. The network automatically changes the difficulty level for bitcoin mining to ensure that a new block can be found every 10 minutes.

Secure-Hash-Algorithm (SHA) The Secure Hash Algorithm or SHA is a family of cryptographic hash functions published by the National Institute of Standards and Technology (NIST).

Sharding Originally, this meant the distributed storage of large amounts of data on different databases. In relation to the blockchain, sharding means no longer having all nodes process a transaction, but only assigning this task to a small number of nodes.

Smart Contract Smart contracts are contracts whose terms are recorded in a computer language instead of legal language. Smart contracts can be executed automatically by a computer system, such as a suitable distributed ledger system.

Software as a Service (SaaS) SaaS is a natural evolution of software. The old model of physically installing software on data centre servers and end-user computers was the only realistic solution for many years.In recent years, a number of developments have led to SaaS becoming mainstream. One factor is bandwidth; the internet is simply faster than it was a decade ago and access is more widely available. Another important factor is the growing acceptance of cloud computing for business use.

Participant An actor who can access the general ledger: Read records or add records.

Timelock A timelock is a type of encumbrance that restricts the issuance of a bitcoin until a certain future time or block amount. Timelocks are an integral part of many Bitcoin contracts, including payment channels and hash timelock contracts.

Transactions Transmission of digital monetary units as well as information.

Transaction costs See fee. Immutability The property of data to be resistant to change. Immutable data is considered "set in stone" and remains unchanged for the rest of time. Data can be functionally immutable, meaning that it is possible to change it, but it would require excessive resources to do so.

Wallet Software that contains all your Bitcoin addresses and secret keys. Use it to send, receive and store your Bitcoin.